Birmingham's working horses

It's the very last sale in Birmingham's Horse Fair - and Brummies turn out in their hundreds to share the passing of an era. (1911).
(Birmingham Central Library)

A CENTURY OF HORSEPOWER

Birmingham's working horses

ON ROAD, RAIL & CANAL

BRYAN HOLDEN

Captions and additional material
by Peter Ricketts

Barbryn Press Limited
BIRMINGHAM

Books by Bryan Holden

PORTRAITS OF 'KINGS'
PORTRAITS OF 'CASTLES'
PORTRAITS OF WESTERN 4-6-OS
(with Kenneth Leech)
BEST OF 'STEAM SCENE'
(Scripts of the BBC radio series)
THE LONG HAUL - the Story
of the Railway Horse)

FOR MY WIFE, BARBARA
WITH LOVE

ISBN 0 906160 08 1

First published April 1989 by
Barbryn Press Ltd, 75 Harborne Road
Edgbaston, Birmingham B15 3BU.

Design: David Warmington Artwork: Geoff Weaver Print Production: Mike Jones

Typeset by Repro Communications Ltd, Halesowen, West Midlands B63 3SW
Printed by Everyware Oscott Print Ltd, Kings Heath, Birmingham B14

Acknowledgements

This book could not have been written without constant help and encouragement. My sincere thanks, therefore, to Peter Ricketts for writing the captions to the photographs and giving me unstinting editorial assistance. To Maureen Messent, for her series of articles which drew a host of recollections and vintage photographs from 'Evening Mail' readers. And to the 'Birmingham Post and Mail' group of newspapers for marketing services.

Thanks to Councillor Bryan Bird, Chairman of Birmingham Leisure Services Committee, for helping to put the Birmingham Horse Parade back on the streets; and to Mike Smith and David Hill, of the Department of Recreation and Community Services for practical assistance.

I am indebted to Stephen Price and staff at the Birmingham Museums and Art Gallery for research assistance. Likewise, to the Central Reference Library's Local Studies Department.

My thanks also for providing valuable photographs and archival material, to: R Galliers, Asst. City Engineer; Dr Bryan Davies, Chief Veterinary Officer; Chief Inspector Turner, West Midlands Constabulary Mounted Division; Miss D Williams, Severn Trent Water; Mr Bernard Painter, W H Painter & Sons Ltd., and Mr David White, Birmingham Dairies Ltd.

I am grateful to British Rail for access to Snow Hill Station shunt horse stables; Peter Anderson, for information on horsebrasses; Albert Judd for personal reminiscences, and David Hope for loaning postcards from his vintage collection.

For their work on the book itself, I thank David Warmington, book design; Geoff Weaver, typography and artwork, Mike Jones, print production; and Barbara Cleverley and Jane Hunt for typing the manuscript in its many stages.

And not least, my heartfelt gratitude to the many 'Evening Mail' readers, and others, who have loaned photographs and given their time to help in so many ways to make this book a lasting tribute to Birmingham's working horses.

Foreword

Since Birmingham was first mentioned in the Domesday Book the work horse has played an integral part in the economic, industrial and commercial life of our City.

Many of us will recall the familiar sight of the working horse driven through the City by tradesmen of various calling - the butcher, the baker, I don't remember a candlestick maker, but I certainly remember the Co-op milkman with his giant horse with the most gentle of temperaments.

And who can forget the traditions of May Day when the horses were beautifully decorated with brightly coloured plumes and highly polished brasses?

This book is written in honour of all those horses which, before today's high speed mechanical age, were the most important form of transport. I know we will never return to the olden days but wouldn't it be a sad loss to forget the role that the work horse played in our City's growth?

Councillor Bryan Bird
Chairman
Leisure Services Committee

Contents

Introduction

The idea of writing a book in celebration of the working horses of my native city has long been in my mind. And certainly there could be no more apposite time to do so than now, in Birmingham's centenary year.

The difficulty has been to compress a wealth of collected material. So I have pared the narrative and concentrated largely on a pictorial description.

But it would not have been so comprehensive and intimate without the dozens of box-'Brownie' snapshots which have lain around dog-eared and fading for half a lifetime.

Hopefully, there may well be a working future for horses in Birmingham. There are, I believe, tentative plans for a Museum of the Working Horse. And the city's Chief Veterinary Officer, Dr Bryan Davies, has indicated he would like to see horses working in our parks and in a number of other selective locations.

What a superb attraction horse buses would be taking sight-seers at a leisurely trot around the highspots of our city, both old and new.

Certainly the work-horses of old have a special place in the hearts of the many who still recall them.

Maybe, *'Birmingham's Working Horses'* will go some way to repaying a long-standing debt of gratitude to both man and beast.

Bryan Holden
Solihull.
February 1989.

BIRMINGHAM'S EARLY DAYS

1 WORKING HORSES 1

Our story of the centenary of the horse in Birmingham really needs to start further back than 1889.

The increase in horsed transport had begun fifty years before, starting in the relatively brief years of the coaching era.

But the coaching days were virtually over as soon as the first public steam railway got underway. This was in 1825, when George Stephenson's 'Locomotion' literally blew the whistle on its rivals.

The death knell of the road coach in Birmingham was likewise sounded, in 1838, on the arrival from London of the first train in the newly opened Curzon Street station.

Booking offices and stables of the numerous coach and omnibus companies were in the principal inns of the town. Here, travellers and coachmen could be comfortably lodged and fed, and their horses rested.

Unfortunately, none of these inns has survived the changing face of Birmingham, although the Woolpack Hotel in Moor Street hung on until about 1950. However, the whereabouts of many of these old coaching inns can be traced by reference to Osborne's map of Birmingham as it was in 1840.

Hagley Road was one of Birmingham's most prestigious thoroughfares, lined with the huge Victorian homes of the city's upper crust. Nowadays a good number of the houses serve as hotels. So great was the power of the residents that they were able to stave off for years the threat of the clanking ugly electric trams. So this scene of horse buses plying up to the Kings Head at Bearwood was preserved for a while.
(David Hope Collection)

The street looks very much the same today - only the traffic has changed! This was Corporation Street before the internal combustion engine brought its fumes, and congestion.
(David Hope Collection)

A third horse was hitched up to help this Birmingham two-horse bus up the hill at Tyseley. And hardy passengers who braved the weather could sit on top in the "garden" seats.
(Patrick Whitehouse Collection)

A picture that will summon up nostalgic memories for many older Brummies - Queen's Drive at the old New Street station. At the head of the cab rank is a LNWR station bus. The photograph was taken in 1885.
(National Railway Museum)

Villa Road, Handsworth, had a pleasant rural calm to it in the days when the locals took the horsebus into town. The electric tram car, however, was soon to dominate all the major routes to the ever-spreading suburbs.
(David Hope Collection)

There were coaching inns within yards of each other in Birmingham's High Street on the site now occupied by Marks and Spencers and the futuristic Pavilions shopping centre.

They were, of course, coaching inns, stop-over places for the hardy folk who dared to face the privations of long distance coach-horse travel and defy the Dick Turpins of the day.

The three were the Castle, the George, and the Albion, all now sadly gone forever like all the similar coaching inns dotted around the town centre.

Down in Digbeth there was another George and the White Hart, both on the left hand side after passing down the Bull Ring. Not only has the Hen and Chickens vanished, but so has the road it stood in - Worcester Street behind the old Market Hall.

In Bull Street there was the Saracens Head and older readers may remember the Stork at the corner of Old Square and Lower Priory.

The Union was naturally enough in Union Street and the Nelson opposite Nelson's statue just off the Bull Ring in Spittal Street.

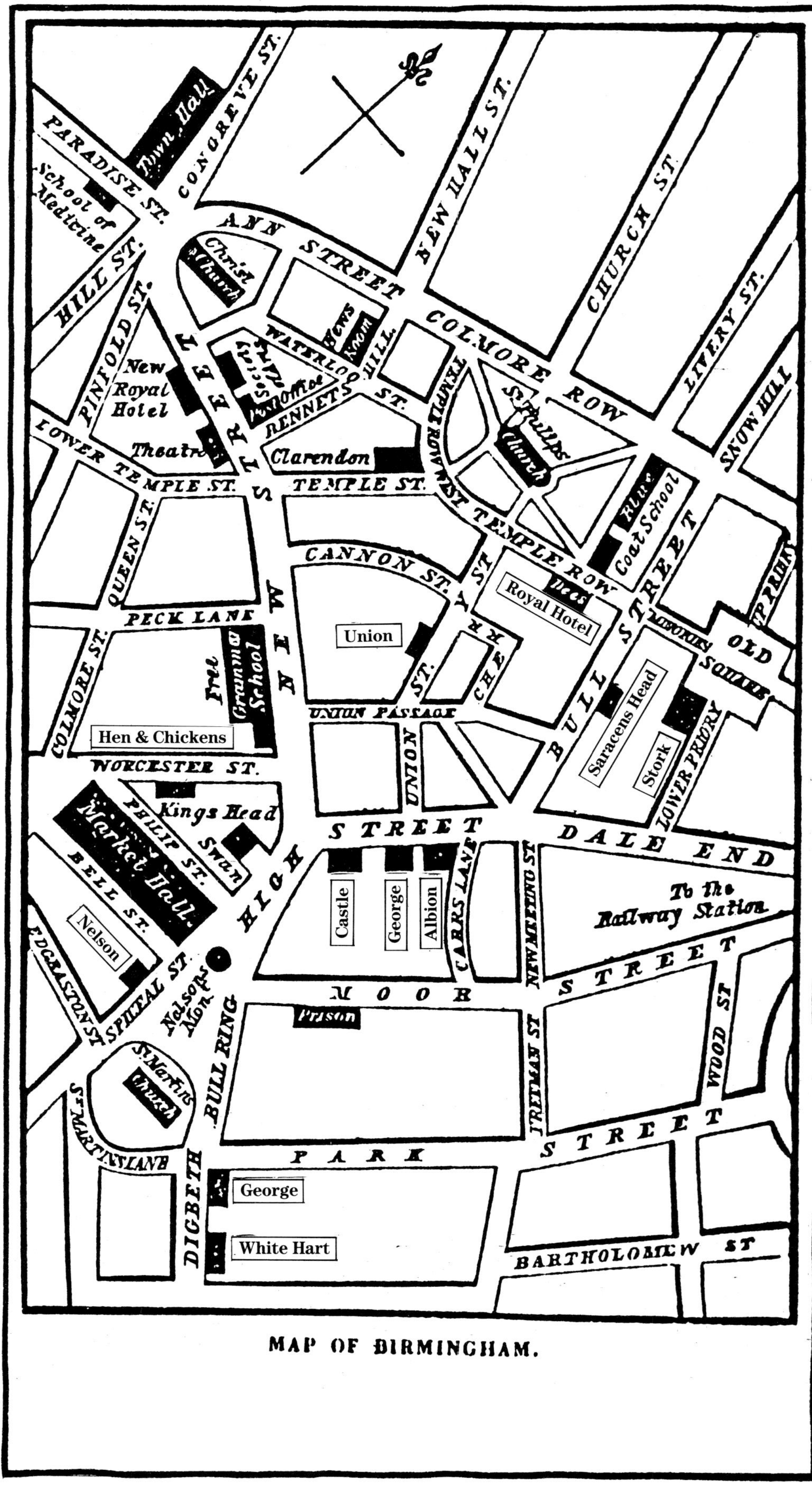

MAP OF DIRMINGHAM.

The coming and going of coaches and omnibuses in the narrow, crowded streets of the town centre must surely have been a stirring sight. In those days coaches from Birmingham ran to all points of the compass; long-distance to Aberystwyth, Bangor, Bristol, Liverpool, Manchester - more than sixty different locations.

The Midland Omnibus Company served the outlying towns of Wolverhampton, Stourbridge, Dudley, West Bromwich, Willenhall, and points westward. The company had offices and starting places in five inns of the town, including the Red Lion (High Street) and the Lamp Tavern (Bull Street).

The coming of the railway also finished the Royal Mail coaches, and the big cut-back in horse usage not only hit the coach owners in the pocket, but threw out of work hundreds of ostlers, farriers, forage merchants, and many others concerned with the welfare of horses.

In a hopeless attempt to get on level terms with the railways, horse coach speeds were improved to an average on main roads of ten miles per hour. But it was a hopeless battle. Flesh and blood was no match for the iron horse, and within a few years the colourful era of the long-distance road coach had gone forever.

A picture that captures all the elegance and tranquillity of a past age. The year in 1912 and a horse bus makes its impressive way along the Stratford Road at Sparkbrook.

NECHELLS
C.B.T.Co. L
HOLDERS CELEBRAT
THE LAST HORSE TRAM
NECHELLS
PHOTO BY A TWIGG
BLOOMSBURY ST
SEP 30.06

The last Birmingham horse tram, 30 September 1906. It plied between the city centre and Nechells and the bowler-hatted driver and the flat-capped conductor found time to pose for the Victorian photographer to set up his equipment which was far removed from today's instant technology. And of course, it was inevitable that it drew a group of children in their smocks and floppy hats to watch the excitement.
(Birmingham Transport Museum)

However, in a growing Birmingham, the horse-taxi, the Hackney coach, was still far short of its Victorian hey-day. The first Hackney cabs appeared in 1832. They were two-wheelers licenced for hire, and drawn by horses with a spectacular elevated trot.

Also in the 1830s, the 2-wheel Hansom cab was introduced. This vehicle had a singular connection with Birmingham; the first of its kind was named after its inventor, Joseph Hansom, the architect of Birmingham Town Hall.

Hackney coach rides were charged by the number of yards distance travelled. For example, the distance from Curzon Street railway station to the Botanical Gardens, Edgbaston, was 4,150 yards, a fare of 2s 6d (12½p). From the Station to the Town Hall, a distance of 1590 yards, the charge was 1s 0d (5p).

The beginning of the Edwardian era saw a rapid decline in the number of horse-drawn vehicles of all types, and by 1909 sixty-eight motor cabs were operating in Birmingham. □

The ladies and children are dressed in their finery for a trip out in the four-wheel family runabout. The driver is Mr James Hunt, who ran his boot and shoe shop in Lozells. Mr P E Madkins, of Alvechurch provides this picture of his great-uncle.

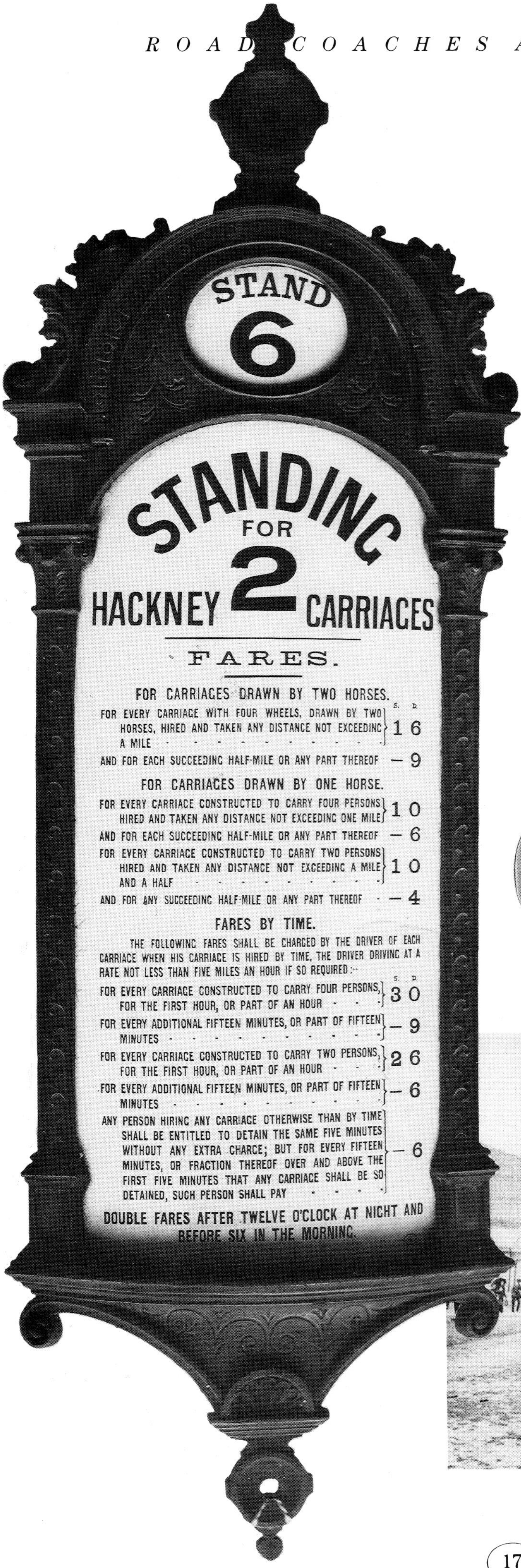

In the early 1900's there were over 60 horse-drawn Hackney carriages operating in Birmingham - and customers were charged according to time or distance and type of carriage and number of horses. Very complicated! Hence signs such as this one.

Commonplace it might have been a few score years ago but now this is a collector's item. It's a brass badge once worn by Hackney carriage driver number 215. The leather thong has a slit so that it could be fixed to a coat button.

(Bryan Holden)

A scene that captures the flavour of the late Victorian era. A Hansom cab driver waits patiently while his horse drinks from a trough in Birmingham's Station Street. In the background looms the vast roof span of the old New Street Station. The picture dates from 1892.

(Birmingham Post & Mail)

Before local government provided comprehensive fire cover, the big industrial companies formed their own brigades. Here are the brave brass-helmeted lads of the Kynoch Works with their engine.
(David Hope Collection)

One of the sights seen in Lozells in the early 1900's was this smart oil tanker done up to look like a steam engine! Mr P E Madkins, of Alvechurch, explains: "It was created in gleaming brass by my grandfather, Henry Josiah Hunt, who sold oil at a penny a pint from his shop in Carpenters Road. It was the only wagon he had, but he painted a "3" on the side to give the impression he had a fleet."

Birmingham's 2 Working Horses

Right to your doorstep

The horses closest to the people of Birmingham were, of course, those making daily deliveries. Milkmen and bread roundsmen were seen by many customers as part of the family, as trusted friends who never let them down.

The horses were regarded as household pets. They came when called by name, always eager for the expected tit-bit - a crust of bread, lumps of sugar, even sweets and chocolate. Children adored the horses, which is why the roundsmen were always dogged by a band of young helpers.

One of Birmingham's best-known dairies was the Midland Counties. Founded by the White brothers, Edwin and James, it set-up in business in 1898 as 'White Brothers Botanical Brewers', making ginger beer.

Here's a smart line-up of horses and milk floats about to begin the daily round. This was the scene at Handsworth Dairies in the mid-twenties.

In 1912 the brothers dropped the 'pop' business, deciding there was more money in delivering fresh milk. The firm moved from Wolverhampton into Aston Brook Street, Birmingham, and became the Midland Counties Dairy Company.

Milk was sold door-to-door from churns on handcarts, but after a few years these were replaced by pony-carts, well-liked for their smart turnout and sturdy Welsh ponies. Realising the affection their customers had for the delivery horses, the dairy capitalised on their popularity and each year put on its own May Day procession.

The milk business also flourished a few miles west of Birmingham, where in March 1910, Carleton White and his brother Val, bought a small milk round, and started with 175 customers.

Like the Whites (no relation) in Aston Brook Street, the emergent Handsworth Dairies was a pioneering family enterprise. The first new customer obtained was a Mrs White. Co-incidentally, she was the mother of Edwin and James White, of the Midland Counties!

Handsworth Dairies also staged its own May Day horse parade. Its horses were top prize-winners at open shows, including Smethwick, Sutton Coldfield, and even as far out as Evesham and Cheltenham.

During the Second World War, in the blitz on Birmingham, in November 1940, the firm's Gooch Street stables suffered a direct hit. Three horses were killed and a number wounded. Two were blown through the stable wall and fell 15 feet into the canal - and survived!

Mr F Bennett, the horsekeeper, was awarded the RSPCA medal for his rescue work.

Early days for the brothers who bought a little dairy in Handworth and built it up to be one of the biggest in Birmingham. Val and Carleton White at Woodville Cottage in Holyhead Road with their first pony, 'Old Dick.' The picture was taken in 1910 and 'Old Dick' was to give loyal service for 16 years.

Mr Knibb was known as the Midnight Milkman when he did his round at Ward End in the 'thirties. He was so keen on a smart turn-out that he was rarely ready to start before 11am and consequently would be seen driving home late at night. The story is recalled by his son Mr H Knibb of Kingshurst.

The state of the stables was a good measure of a dairy's standards, and here's one of the best - clean, orderly, and happy horses with their tails neatly cropped. Handsworth Dairies (Circa 1930)

Albert Judd went out on his rounds from Vauxhall Dairies for 42 years, and here he is in those days introducing the children to Bonita while delivering milk in Bowyer Road. "Juddy" has entertained the crowds driving and chatting his way round the first two reborn Horse Parades in Cannon Hill Park. "They'll have to shoot me to stop me doing this," he told onlookers.

Birmingham's most famous milkman was Albert Edward Judd of Castle Bromwich. He served as a roundsman with the Birmingham Co-operative Society for 42 years. He started with the Co-op in 1940, and being unfit for military service, joined the Air Raid Precautions ranks as a first-aider.

On the same night as the Gooch Street stables were destroyed, 'Juddy' wrote his name in the history books. He was on duty in Coventry Road, Small Heath, when the raid began, and immediately rushed to Whitmore Road stables to see if the horses were all right. On the way, he commandeered a bus to take injured people to hospital, hanging on outside the cab shouting directions to the driver.

He then went on to Whitmore Road where he found the stables had been hit. The horses were in a terrible state, dead and injured animals hopelessly trapped under tons of debris. Juddy (all seventen stone of him) climbed in through a broken window and managed to release the uninjured horses and set them free in a nearby park. He was later presented with an award and commended for his brave action by Lord Dudley.

In the great days of the horse, the Co-op would bring not only milk and bread to your door, but coal, laundry; fresh fruit and vegetables, too. With so many departments and services, and over 700 horses, the Society was able to put on the most spectacular parade tableaux.

Birmingham's best known horse, when annual shows brought colour and excitement to many areas of the City, was 'Verity' the pride of Scribbans' bread delivery stable. But, in truth, Verity was a cut above the ordinary workhorse and did not do his fair share of the daily grind. He was more at home on parade at the big shows, winning prizes wherever he appeared.

But although the big names took pride of place, many folk still recall with affection the smaller traders. The local butcher's gig; the paraffin-seller's cart, top-heavy with all sorts of pots and pans, looking like a horse-drawn armadillo. The rag and bone man's flea-bitten nag, a bucket of fungus-ridden goldfish swaying between the back wheels. Happy Days! □

Mr Norman Watts of Stechford has this reminder of the days in the late 'twenties when he put in many long hours delivering bread for H Joyner of Stechford. "We even had a rota to go in on Sundays and Bank Holidays to feed and water the firm's four horses," he writes.

Mr George Bailey carries the basket in one hand, steadies his horse Bullet with his other, and is about to "deliver their daily bread" to customers of Harding's famous bakery. The picture was taken in 1920.

Wacaden (Wathes, Cattell and Gurden) milk was the favourite for thousands of Brummies and among those delivering it - for 24 years - was Mr Len Brown of Kings Heath.

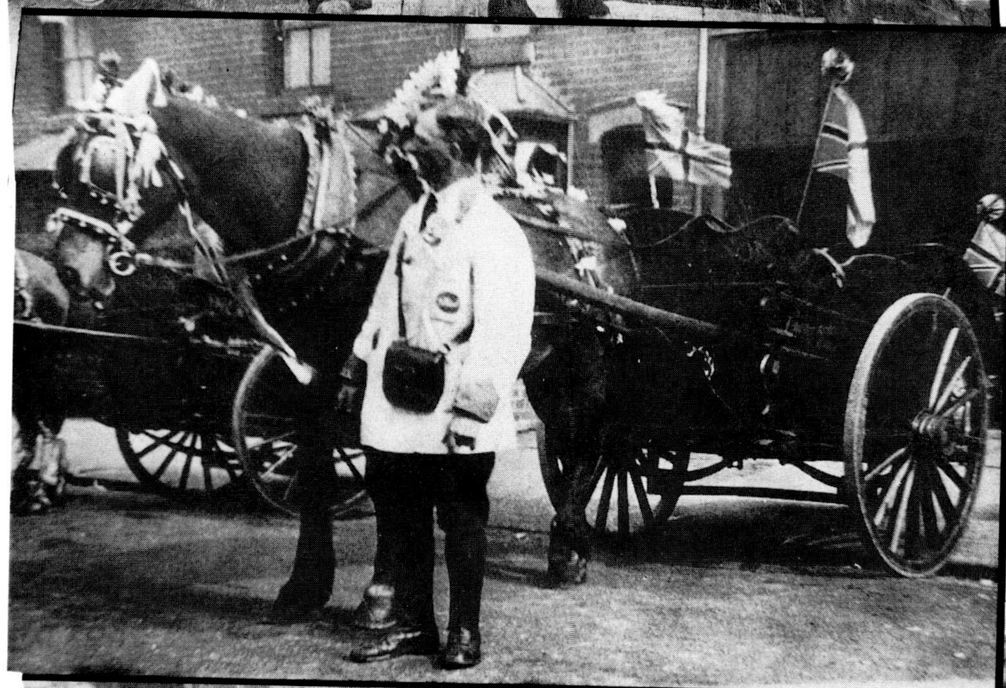

"My dad was a milkman for James Holt of Aston from about 1923 until he died suddenly in 1942," writes Mrs K Taylor of Birmingham B28. She has this picture of her father, known to all as George, at work in Bordesley Green. "The milk cart did not provide seating and dad had to stand the whole way," says Mrs Taylor. "And after the day was over he had to clean the churns, yard, and stables, but he loved it and lived for his horses."

Mr A E Watkins, of Birmingham 32, treasures this photograph of his father holding the reins of a Birmingham Co-op milk float. "My father was one of the first seven roundsmen to be employed by the Co-op in 1920 when milk was fourpence a pint and it was served with a ladle," says Mr Watkins.

Dinkie was a lovable rascal of a horse and a favourite with the roundsmen of Rowland's Bakery. Once in his early days the breadman returned from serving a customer in Kingstanding to find that Dinkie had disappeared, cart and all. He rang the bakery to give them the news and was told "No wonder you can't find her - she's back here!" Recalling the story, Mr C J Rooke of Sheldon points out that Dinkie had travelled all the way up Birchfield Road untended and without a mishap. He later took on the wily horse himself and is seen with her here, dressed for a local show.

After working with horses as a Birmingham Co-op coalman for 22 years, the time comes for Fred Gibbs to lead his faithful old friend Carol away from the wharf for the last time. Now his daughter Mrs K E Austin of Northfield, treasures the photograph as a memento of a past era.

Mr Norman Watts of Stechford treasures this photograph of his father in the days when he was a traveller for Lyons tea. The vans were loaded up outside the Lyons depot in James Watt Street and off would go Mr Watts senior, perhaps on a three-day journey, staying overnight at inns along the way.

Right: Harness makers were worth their weight in solid leather if not gold. Here is Freddie Hunt at Handsworth Dairies with a magnificent example of his trade.

The anvil is on the right, the box of tools is to hand, as Handsworth Dairies' farrier Syd Roberts plies his trade.

Frank Summerton is in winning form, taking the cup for the Cleanest and Best Decorated Turn-out. Frank joined the army during the second world war, and afterwards became manager of Handsworth Dairies' South Birmingham branch.
(Birmingham Dairies Ltd)

The war broke up the partnership between Queenie and John Skinner, a bread roundsman at Quinton for C E Powell of Harborne, But they got together again when peace returned and here they are in Harborne High Street. Soon, electric carts pushed Queenie into retirement, but she had the last laugh during the heavy snows of 1947. The new-fangled carts could not get through, but Queenie did - and earned Mr Skinner 50 new regular customers.

Mr R S Pardoe, of Stechford, was a 50p-a-week bread roundsman at 14 and here's his photograph to prove it! The war had just started and the reason for Mr Pardoe's rapid advancement at a bakery firm called J E House Ltd of Small Heath was because many of the older men had been called up.

Far right:
Man and horse in perfect harmony... this picture captures it all as H Nevett, of Yardley, delivers bread for the Co-op in Mapleton Road, Hall Green, in 1935.

Mr Norman Teece poses while a customer takes his photograph in Reynolds Road, Handsworth. The year was 1949 when Mr Teece worked for Handsworth Dairies. The name of the horse was Chestnut Bob.

One of the worst winters on record brought chaos to Birmingham's streets in 1947 - but the milk still got through. Here's one way the Co-op delivery men coped - by building a makeshift sledge from stacking platforms usually used inside the dairy.

After the snows in 1947 came the great thaw bringing still more problems for the milkmen... and in Tame Road, Witton, a Co-op customer waits hopefully to catch the special delivery as it sails through the air!

Birmingham's Working for the Corporation

Working Horses

Public Works covered everything from repairing holes in the roads, street cleaning and lighting; to refuse removal, laying paving stones and cobbled setts, and the maintenance of sewers and drains.

In Birmingham's early years, such mundane tasks needed an army of men and horses working around the clock.

In 1891, the Corporation had 289 horses in its stables, of which almost one hundred belonged to the Public Works Department.

The size of the horse complement led the War Office to enquire in 1898

Horses still played a big role in keeping the city running, even in the 1940s. Here, Mr Harry Fryer is seen with Strawberry Roan, a Shire of that colour, at Calthorpe Park in 1947.
The cart is typical of its day, sturdy oak frame and wooden-spoked wheels protected by a strong iron "tyre".
(Birmingham Environmental Services)

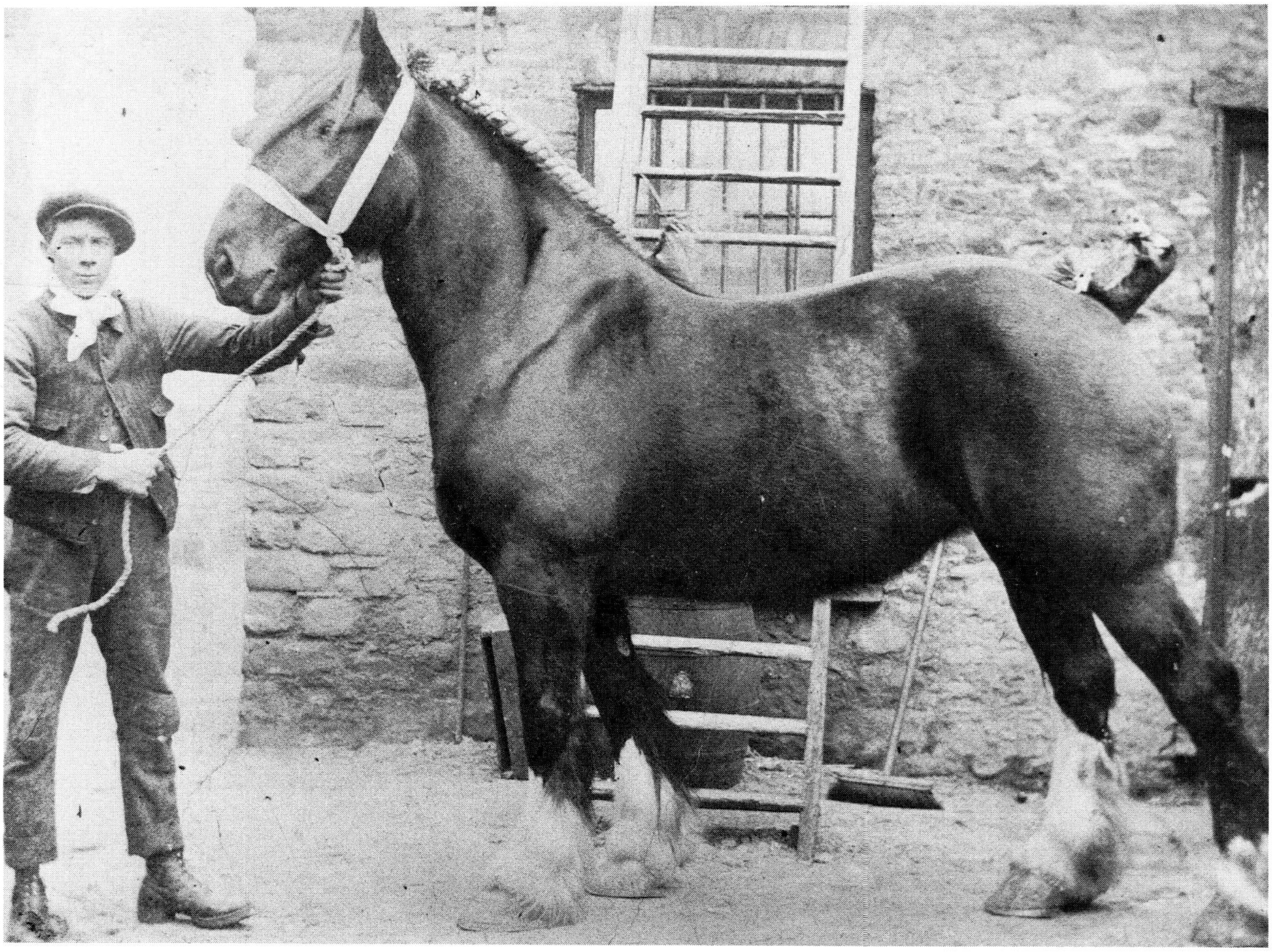

Many of Birmingham's best working horses before the first world war came from the stables of Mr Fred Grindle at Hanham Green Farm, Bristol, "the biggest horse dealer in the west." Here a groom displays a fine specimen at the Bristol centre. Note the closely docked tail, a practice which became illegal.
(Birmingham Environmental Services)

whether any of the horses might be registered for military service in the impending war in South Africa.

Not surprisingly, the Corporation wasn't enthusiastic about the idea and said it regretted that few, if any, of its animals were suitable.

When the 1914 war began the Public Works Department owned 352 horses and this time were given no choice - Kitchener commandeered the best animals and sent them to France.

The next largest public services owner was the Refuse Department with 279 horses.

The welfare of the horses was necessarily of the utmost concern. A Corporation sub-committee supervised the horse department, and was responsible for the appointment of a veterinary surgeon.

Eventually, a 76-acre farm was acquired in California, in south-west Birmingham, and a horsekeeper lived there until 1947. The farm was used as a rest and convalescence home for Corporation horses and helped to feed them with a hay crop.

The highlight of the drivers' and the stablemen's year was a string of annual horse shows leading up to May Day. Many horses took part in the smaller shows, but the pick of the stable competed in the biggest event, the Birmingham Horse Parade, in Calthorpe Park, Balsall Heath.

A fine bay Shire employed by Birmingham Public Works Department. This one is called Punch and proudly showing him at Kings Heath Park in 1931 is Mr Sidney Thatcher.

(Birmingham Environmental Services)

Mr J Histon with his entry for the Birmingham Horse Parade, Calthorpe Park, in 1930. Mr Histon worked with horses for the Public Works Department for 40 years at Coventry Road and Kings Heath.

(Birmingham Environmental Services)

The City of Birmingham's coat-of-arms can be seen on the blinkers of this dapple grey Shire, photographed at the Calthorpe Park May Day show in 1937.

(Birmingham Environmental Services)

When horse power was at its peak, Birmingham Corporation kept hundreds of horses in stables throughout the city. The picture below was taken at the Public Works yard in Kings Heath and shows Mr M Cooke, of Weoley Castle, who worked there and who provides the photograph.

A steep roadway led to the rubbish hoppers belonging to Birmingham Salvage Department at Montague Street and a 'mechanical chain horse' helped the loaded dustcarts up the slope. Here's one attached to the chain and about to start the climb.

(Birmingham Reference Library)

Above: The last of the Birmingham Parks horses... Tommy was Mr Harman's constant workmate for 17 years. He was stabled at Ward End Park, Washwood Heath. The pair are seen together on May Day 1948 after winning third prize in the Birmingham Horse Parade.

Picture from his son, Mr W. Harman of Alum Rock.

Wagons roll... at the Birmingham Horse Parade at Calthorpe Park in 1937. On the right is Mr Harry Fryer with Nigger pulling a Public Works Department cart. The covered wagon, western style, belongs to Mitchells & Butlers Brewery.
(Birmingham Environmental Services)

There is a pile of old horse shoes in the archway at the Holliday Street stables and it is possible that Birmingham Corporation's Mr Sidney Butler had just had this fine bay Shire re-shod before setting off for another day working for Birmingham Corporation, in 1937.
(Birmingham Environmental Services)

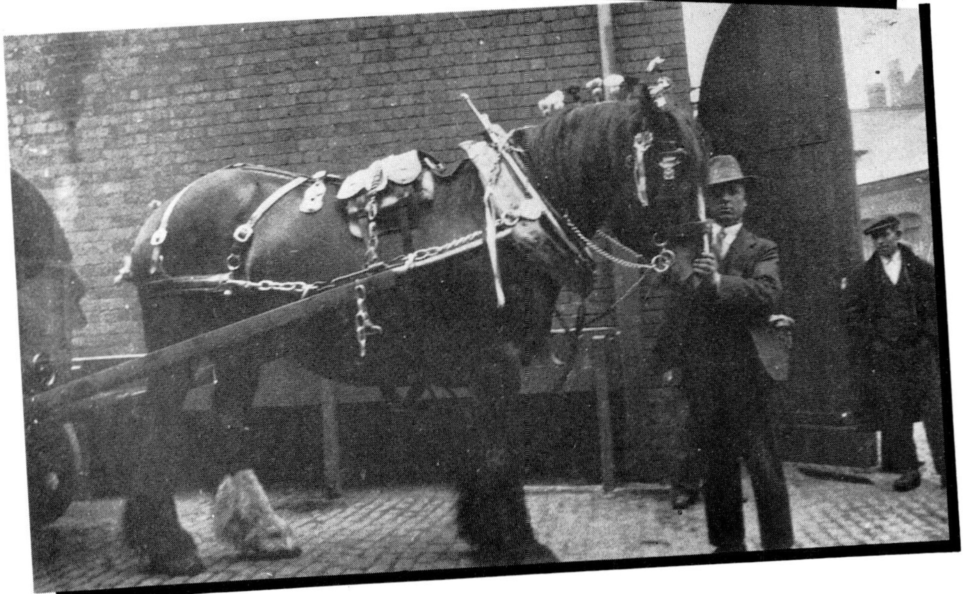

Strangely, neither the name of the horse nor its driver is known - but we know that the onlooker is Birmingham Corporation's Mr Harry Fryer. The picture was taken at Holliday Street in 1937.
(Birmingham Environmental Services)

Stark brickwork, cobbled pavements, and a general air of decay, fail to blur the magnificence of Robin, the brown Shire. And small wonder because he has been groomed until his coat is shining with health, the leather harness gleams, and the brassware has been burnished until it glows like gold. Such was the May Day scene at the Corporation stables at Holliday Street in 1937. The elbow grease proved worthwhile. Mr Howard Oakley (left) took Robin off to the show and won the Birmingham Challenge Cup.

(Birmingham Environmental Services)

Public Works horses were regular prize winners, and before the Second World War, the W.R. Fleetwood Challenge Cup was won outright by the department, with three successive wins.

Three years after their final show appearance, at the Handsworth Show in 1951, the last five Public Works horses were sold out of service. Modernisation had won the day. But thirty-six years later, horses were

Same time, same place, and this time Mr Sidney Nash prepares Tango, a bay Shire, for the 1937 show.

(Birmingham Environmental Services)

In 1975 an official working in Birmingham Engineers Department persuaded a reluctant store-keeper to lend him the key to a vast Victorian horse feed mill in Holliday Street in the city centre.

The mill, built in the 1880's because of the unreliable and generally poor quality of commercially processed feedstuffs, had been unused and locked up for many years ...

...what was found in the old mill was astonishing. All the paraphernalia for the production of horse feed in huge quantities was intact. Great pulley wheels, driving belts, chaffing machines, chopping devices, mixers, the lot.

The official, Mr R Galliers, recalls: "It was a very unusual experience. It was just like the 'Marie Celeste'. Everything was as it was; there was even hay in a wooden wheelbarrow and beans in a hopper."

Fortunately, before the building was cleared, photographs were taken, and we publish several of them here.

(Birmingham Engineers Department)

back in the Corporation's old Holliday Street stables. By then the majority of standings and loose boxes had become store houses, but a couple were in good enough shape to house Viscount, a Shire gelding, and Chester a cross-bred Clydesdale.

From 25 February 1987, the two were on a six-week trial with the Cleansing Department in a far-seeing project initiated by Dr Bryan Davies, Birmingham's Chief Veterinary Officer.

Viscount hauled a rubbish collection cart around the city centre, and literally stopped the traffic, such was the interest he aroused; while Chester pulled a butty-boat on the nearby canal.

Both horses did their jobs well. The horse-drawn rubbish cart was faster than a motor lorry, due, in part, to the fact that the driver wasn't getting in and out of the cab, or continually stopping and starting.

But it remains to be seen whether the City Environmental Services Department, the modern-day successor to the Public Works Department, puts horsepower back onto the streets of Birmingham. □

The City Vet, Dr Bryan Davies (left) checks over Viscount before he sets off on his rubbish-collecting round from the Holliday Street stables in 1987. The horse was the centre of attraction wherever he went and even became a TV star with an appearance on Blue Peter! Holding the reins is Donald Knight of the Veterinary Section.
(Bryan Holden)

Below:
Move over, we've got more horsepower . . . cars fill the courtyard at the old Public Works stables in Holliday Street in 1987. But horses were still present — including Viscount and Chester on a six week trial working a rubbish cart and a canal butty boat.
(Bryan Holden)

Birmingham's

The city's rural heritage

Working Horses

Find a farm in Birmingham and you deserve a prize, but it was not always so bereft of rural delights.

At the turn of the century Moseley and Kings Heath, Selly Oak and Kings Norton, and Harborne, were still village communities.

Outlying districts, such as Hay Mills and Yardley, Acocks Green and Olton, Hall Green and Shirley, were still green fields where clear streams babbled and children fished for tiddlers.

Farms were smallish family affairs, self sufficient in their own particular way; earning a living as suppliers of milk, eggs, meat and poultry.

Heavy horses were the farm tractors of the day, the jacks-of-all-trades, used

Farmer Frank Smith puts in a good day's work harvesting in 1939 at Frogmore Farm, Fen End, Knowle, with the help of 'Old Bill'.

(Bryan Holden Collection)

Frederick Ford supervises the corn threshing at Dumble Farm, Maxstoke, in the early 1900s.
Picture from his grand-daughter, Mrs P E Knowles of Sheldon.

for ploughing, harrowing and harvesting work, as well as odd-jobbing as the seasons demanded.

With the growth of the railways and the emergence of industry, many country folk left the farms to work in the factories. But some retained their rural skills and became stablemen, carters and horsekeepers.

While not strictly the working horses of the city, farm horses merit inclusion in this tale because they were part and parcel of Birmingham's industrial and commercial scene.

Farmers utilised the services of in-town fodder merchants, saddlers and harness-makers, farriers, wheelwrights and joiners. The whole comprised a many-sided business employing a significant percentage of the city's work force.

Then thousands of acres of arable and grassland were over-run with concrete as the large council estates and private developments spread out into the countryside.

Some farms survived and for years existed as uneasy neighbours to the urban sprawl. One by one they disappeared as demand for land became even more insistent.

Heavy horses, once an everyday sight, became rarities to be admired at special events.

Even today, Birmingham bears testimony to its rural past. For instance,

Needless Alley, in the city centre, still has a kink in it where the ploughmen of old turned their teams of horses.

Birmingham's rural heritage is preserved more noticeably in the names of roads. The list is impressive. In Erdington, for instance, there are Ashold Farm Road, Berwood Farm Road, Pitts Farm, and Faulkners Farm Drive. Northfield has Allens Farm Road, Beech Farm Croft, Hole Farm Road, and Tinkers Farm Road.

Bartley Green has Moat Farm Drive, Quinton has Welsh House Farm Road, and in Pype Hayes there is Woodlands Farm Road. And many many more.

Thus, this book on the working horse in Birmingham, city of a thousand trades, would be incomplete without looking back to the days when fields were tilled in the city and the noise and smell of factories, although only over the brow of the hill, seemed a thousand miles away.

The photographs in this chapter are a tribute to the farming community that lived and worked in Birmingham. An era passed but not quite forgotten. □

There were many flour mills scattered throughout the Birmingham area and this delightful rural scene was at the Olton Mill. An essential part of the business was the horse and dray to bring the corn in and take the flour out.

(David Hope Collection)

A setting that has the hallmarks of a Constable painting - verdant trees silhouetted in a silver lake and on the winding path a carter and his three-horse team wend their way homeward. The picture was taken at Coston, Barnt Green.

(David Hope Collection)

The Coton family wrested a living from the land at Yew Tree Farm, Solihull, for many decades stretching from the last century - and these magnificent photographs show how their daily tasks were closely tied to the power of the horse.

A young Abe Coton outside Yew Tree Farm, to which he was to devote his life. Here he is in 1904 in the arms of his farming father, Ernest Coton. Mrs Coton looks after Abe's sister Ann.

(Mrs. D. Mason Collection)

Abe Coton gathers in the corn at Yew Tree Farm. The original farmhouse was knocked down and replaced in the late 'thirties. This house still stands, near to Damson Parkway.

(Mrs. D. Mason Collection)

Dinah Coton married John Warner and they lived in Lugtrout Lane, Elmdon Heath, in a house that still exists near the junction with Damson Parkway. Here the couple pose with daughter Violet in the family runabout.

(Mrs. D. Mason Collection)

The year is about 1930 and Abe is busy harvesting the mangold crop at Yew Tree Farm. He's the one with the dog. Also in the picture are the farm owner, Fred Lines (centre), and Bert Davies.

(Mrs. D. Mason Collection)

Mr F Hawker of Solihull treasures this picture of his father with horses Dolly and Mary harrowing the land at Dickens Grove Farm, Copt Heath, in the 1930's.

Fred Barnett mowing the grass on a farm he managed in the 'thirties'. The houses in the background are in Wagon Lane, Sheldon, and the Good Companions pub now stands on this site.

Picture from Mr Barnett's old friend Bill Chambers of Shirley.

Horses are still used by the Forestry Commission to work in woodlands which are inaccessible to tractors. This vintage photograph is thought to have been taken on the Lickey Hills.
(David Hope Collection)

Birmingham Co-op had many farms in 1950. The picture shows haymaking in progress at one of them at Lea Marston.
(Central Midland Co-operative Society Ltd)

If ever you ask author Bryan Holden how his great interest in horses came about, he's likely to show you this photograph! It shows him as a lad of eight aboard a cart at Frogmore Farm at Fen End, near Knowle. He would spend school holidays helping out there and, whenever possible, thumb a lift on the cart pulled by this lovely old horse called Bill.
(Mrs Elsie Westwood)

BIRMINGHAM'S

5 5

The railway workhorses

WORKING HORSES

The once familiar sight of the railway dray-horse, huge, docile and ever-patient, has long since gone. Gradually, it was replaced by the mechanical horse, or 'Cob', which could easily be adapted into a shaftless cart, float or lorry.

The seven-horse power Cob needed neither stable, nosebag, nor ostler. It could be put into the garage and forgotten. But a horse needed feeding, watering, grooming and exercising seven days a week.

Railway horses worked in Birmingham until the late 1950s. Heavy horses,

A working day at the GWR goods yard at Hockley in 1921.
(British Rail)

Jack Jackson, one of the last two railway horse drivers at the Hockley Goods Depot, takes the reins of a goods wagon there. At the head is his old friend William Steddan. Picture supplied by E Dawes of Birmingham 21.

mainly Shire type, were used for goods cartage and shunting, and half-heavies, or vanners, for fast parcels and passenger luggage.

The main stables for Great Western Railway horses were at Hockley, with smaller establishments at Moor Street, Bordesley, Small Heath and Tyseley. Shunt horses laboured in the smoke-ridden tunnel sidings at Snow Hill.

Lawley Street goods station, and nearby Curzon Street, accommodated the bulk of the London, Midland and Scottish Railway horses.Other LMS horses were stabled in the former Midland Railway goods yard in Holliday Street; at Monument Lane, Kings Heath, and other suburban locations.

On joining the railway, a youngster had to be seventeen before he could transfer to the loco department. Many would-be engine drivers found themselves working as van boys, or chain horse lads in the horse department, for a couple of years.

Van horse boys were nicknamed 'tailboard monkeys', because they rode in the back amongst the parcels, swinging in and out of the van at the end of a rope.

Chain horse boys helped pull loaded drays up steep inclines by hooking their horses in front of the cart horse, thus doubling the available horse power.

At the height of the soft-fruit season, trains from Evesham running into Moor Street were met by convoys of horse-drawn drays ready to rush the fruit into the Birmingham Market. Extra men and horses were brought in from outlying stations.

A similar activity would be taking place a mile or so up the road, in the LMS goods yard at Camp Hill. The race to be first to the market was a neck-to-neck tussle. Not just a commercial necessity, more a matter of personal honour.

It was vital that horses were kept in good health. Railway companies were always most rigorous in looking after their equine workforce.

Company inspectors roamed the streets, keeping an eye on man and beast. Seeing that carts were not overloaded. Or attempting too steep an incline without a chain horse.

Even drinking water was strictly vetted. No railway horse was allowed to drink from a public trough. A bucket had to be carried under the cart and filled from a freshwater stand-pipe.

It was important, too, to keep horses' feet in good shape. They were inspected daily for signs of injury, corns, and loose shoes. Depending upon the weight of the load, and conditions underfoot, a railway horse, on average, had a new set of shoes every 3 weeks.

During the war many stables were destroyed in air-raids, the biggest hazard being fires caused by incendiaries. In 1941, a bomb hit one of the LMS double-

The last war broke down many traditional barriers and it was not surprising that with the men away in the forces, women took over a few of the carting jobs at Snow Hill station. Here is one of them - Miss Edith Weston. With her is fellow carter Mr Davies.

As late as 1949 the Royal Society for the Prevention of Accidents was awarding safe driving diplomas to railwaymen in charge of horse-drawn vehicles. This one was won by Leonard Lightwood of the London Midland Region of British Rail. And to impress on him that he had won something very special, it was signed by no fewer than four dignitaries!

THE ROYAL SOCIETY FOR THE PREVENTION OF ACCIDENTS

DIPLOMA

FOR THE DRIVER OF A HORSE DRAWN VEHICLE

AWARDED TO LEONARD LIGHTWOOD

OF BRITISH RAILWAYS - LONDON MIDLAND REGION

FOR HAVING BEEN FREE FROM ACCIDENTS FOR WHICH HE WAS IN ANY WAY BLAMEWORTHY, WHILST DRIVING A VEHICLE DURING THE YEAR ENDING 31ST DECEMBER, 1949

PRESIDENT

CHAIRMAN NATIONAL SAFETY AWARDS COMMITTEE

DIRECTOR GENERAL

SECRETARY Helen Sutherland

THE ROYAL SOCIETY FOR THE PREVENTION OF ACCIDENTS

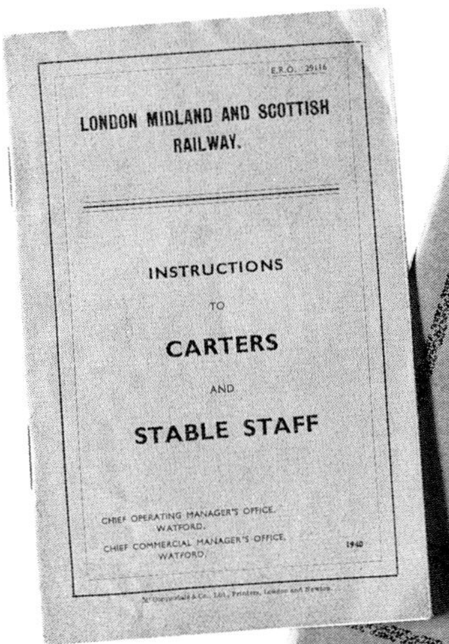

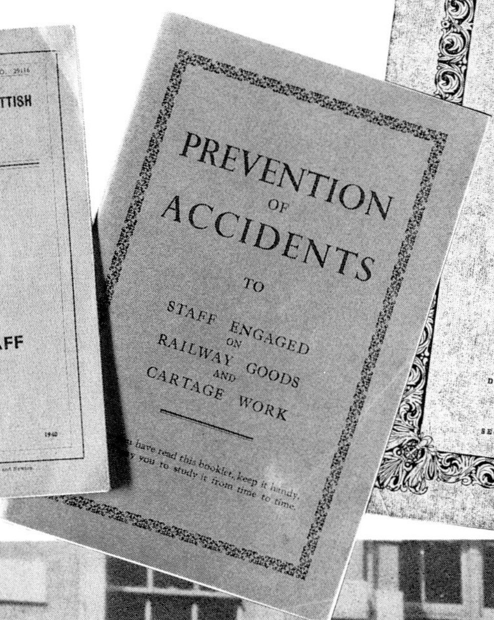

The railway solved the problem of hauling trains up steep slopes by employing "bank engines" to add extra pull. So it was logical to do the same when their horses were faced with the same task - they added an extra horse! Here is a "chain horse," as it was known, helping a horse and cart up the slope out of the GWR goods station at Hockley, Birmingham. At the top it would be unhooked and returned to the bottom for the next cart to come off the weighbridge.

The arrival of petrol driven power spelled redundancy for the "chain horse." Here's a three-wheel runabout, registration number 033 OL, giving a helping haul to a two-horse cart.

The Victorian railway pioneers always planned on a grandiose scale and when stables were needed by the LNWR in Birmingham they built them in the manner of a fortress! Here they are, dominating the skyline and the nearby houses at the corner of Vincent Street and Sheepcote Street in the centre of the city. The stables were also linked to a canal wharf for easy loading and unloading of goods.

They were used as a Public Works storage depot when this photograph was taken in 1986.

(Bryan Holden)

The horsekeeper at the LNWR stables was provided with this house so that he could "live on the job" and keep a round the clock watch on his charges. The stables adjoin on the right.

(Bryan Holden)

decker stables in Suffolk Street. It passed between two horses on the top floor and two below, and embedded itself in the floor, but luckily failed to explode.

About 150 horses were taken from the stables and corralled under a nearby aqueduct while the bomb was being defused.

The highlights of the carters' year were the May Day parades and stable competitions staged by the railways and the RSPCA.

The finalists were chosen by a process of elimination over a period of several weeks, points being awarded for conditon of stable, horses and harness.

Allowance was made for harness showing signs of reasonable wear, but cleanliness was all important. Harness had to be fitted correctly and attached properly to the vehicle.

An important event on the railways of old was the annual horse parade. Keen competition was fun for the railwaymen and it encouraged them to keep their charges in first class condition. And if at the end of the day they could hang a certifcate on the parlour wall, then the whole thing had been well worthwhile. Here are three examples of the highly prized awards.

(Bryan Holden)

Mr A. Haycock worked on the railway at Curzon Street for 48 years and for 11 of those years delivered goods by horse and dray. Every year he would dress the horses up for the big annual shows. Here he is with Entry No 177 Lightwoods Park, Bearwood.

Captain Milne, a retired army captain became a legendary figure just over a century ago, for the way he cared for the hundreds of horses owned by the Great Western Railway.

Indeed, it could be said that Captain Milne created his own "army" of carters, farriers, harness makers and stable hands, all obeying a strict code of conduct that decreed that the well being of the horse was paramount.

The result was that the GWR stables were considered to be as good as any in Britain.

The captain held one of the key posts in the company because, if the horses failed, the freight carrying side of the business would quickly come to a halt.☐

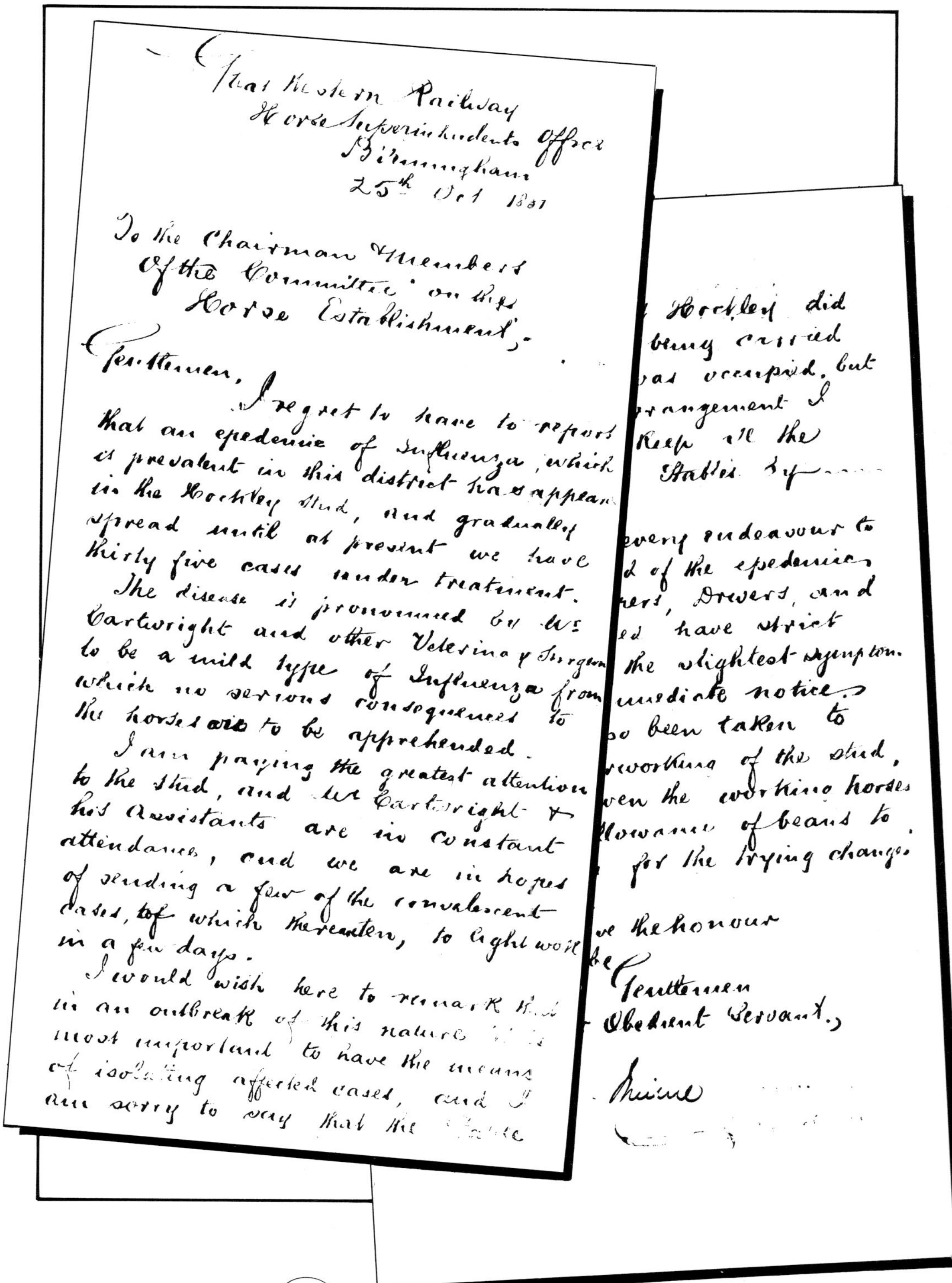

Great Western Railway
Horse Superintendents Office
Birmingham
25th Oct 1881

To the Chairman & Members
Of the Committee on the
Horse Establishment,

Gentlemen,

I regret to have to report that an epidemic of Influenza, which is prevalent in this district has appeared in the Hockley Stud, and gradually spread until at present we have thirty five cases under treatment.

The disease is pronounced by Mr Cartwright and other Veterinary Surgeons to be a mild type of Influenza from which no serious consequences to the horses are to be apprehended.

I am paying the greatest attention to the stud, and Mr Cartwright & his Assistants are in constant attendance, and we are in hopes of sending a few of the convalescent cases, of which thereafter, to light work in a few days.

I would wish here to remark that in an outbreak of this nature it is most important to have the means of isolating affected cases, and I am sorry to say that the Stable

Two fascinating letters written by Captain Milne from his headquarters in Birmingham describe in detail the effect of an outbreak of influenza among the horses kept at the GWR stables at Hockley.

One of them is to the chairman himself, the renowned and much respected Sir Daniel Gooch.

But Milne tells of the outbreak in his first letter, dated 25 October, 1881, addressed to "The chairman and members of the committee on the Horse Establishment."

He states that 35 of the animals had the disease, then goes on to display his remarkably advanced knowledge of medicine... "In an outbreak of this nature it is most important to have the means of isolating affected cases, and I am sorry to say that the stable accommodation at Hockley did not admit of this being carried out as every stall was occupied, but I made the best arrangement I possibly could to keep all the affected horses in stables by themselves."

The letter to Sir Daniel Gooch is dated 1 November, 1881, and states that there were now 63 cases of influenza but of these 33 had been treated successfully and were being given light work.

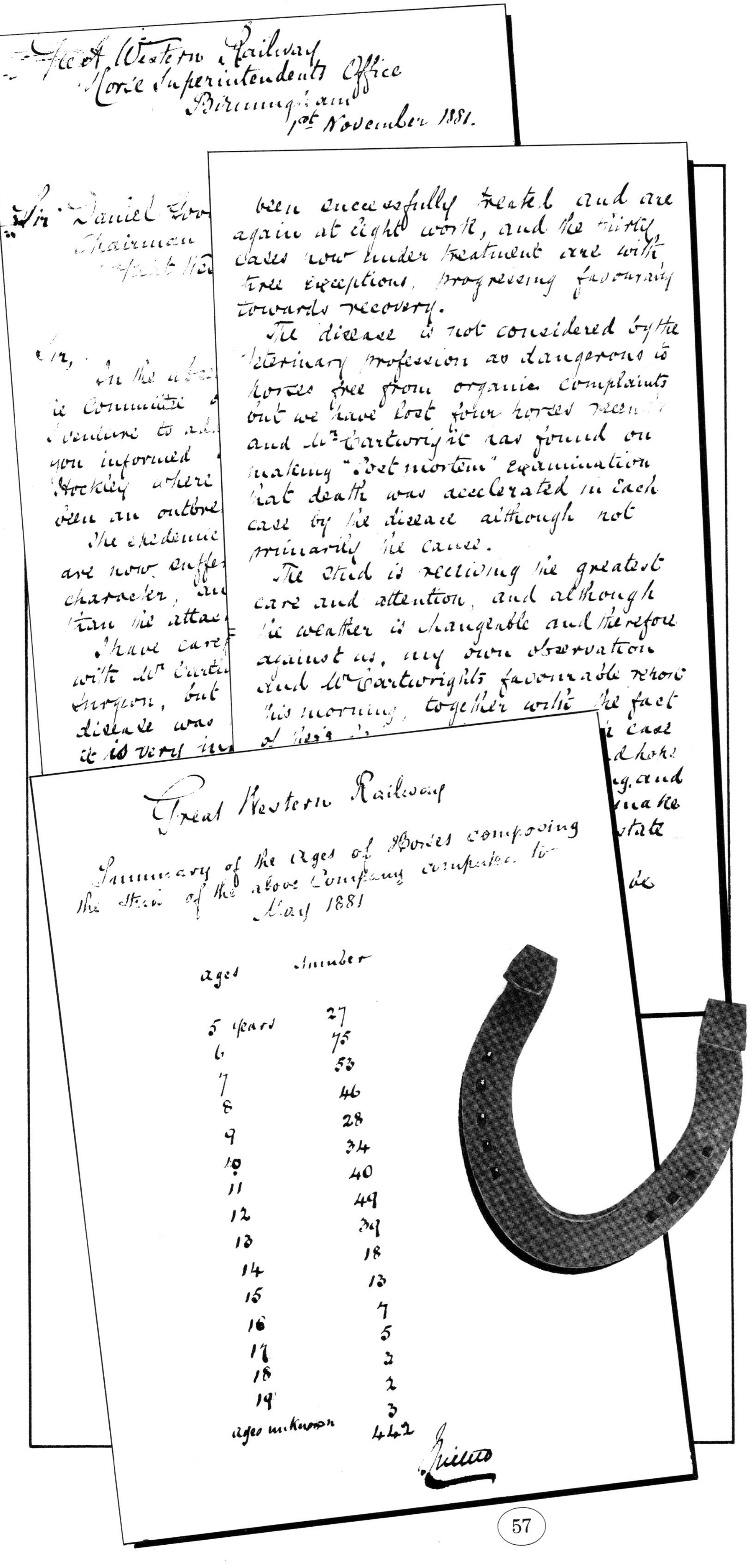

Great Western Railway
Horse Superintendent's Office
Birmingham
1st November 1881.

Sir Daniel Goo[ch]
Chairman

been successfully treated and are again at light work, and the thirty cases now under treatment are with three exceptions, progressing favourably towards recovery.

The disease is not considered by the Veterinary profession as dangerous to horses free from organic complaints but we have lost four horses recently and Mr Cartwright has found on making "Post mortem" examination that death was accelerated in each case by the disease although not primarily the cause.

The stud is receiving the greatest care and attention, and although the weather is changeable and therefore against us, my own observation and Mr Cartwright's favourable report this morning, together with the fact

Great Western Railway

Summary of the Ages of Horses composing the stud of the above Company computed to May 1881

ages	number
5 years	27
6	75
7	53
8	46
9	28
10	34
11	40
12	49
13	39
14	18
15	13
16	7
17	5
18	3
19	2
ages unknown	3
	442

Milne

The captain goes on: "I have carefully gone into the matter with Mr Cartwright, the consulting veterinary surgeon, but we cannot trace how the disease was introduced into our stables. "It is very infectious and prevalent in the district so that it may have been conveyed from other horses in the streets. The drainage of our stables is perfect and there is no local cause to which the outbreak can be attributed."

(Bryan Holden Collection)

Captain Milne kept meticulous records of his stock, their numbers, well being, and location, and all in beautiful copperplate writing. Many of his reports are still in existence. One, dated May, 1881, lists the number of GWR horses in age groups. We learn that of the 442 animals then employed on the railway, no fewer than 90 were over 13 years old, two of them 19 and still working!

It is an indication of the expert care and attention that the Milne regime lavished on them.

M R
M R

This magnificent photograph of Lawley Street goods yard was taken in 1892. The railways showed great ingenuity in designing horse-drawn wagons to suit every need. In the centre a horse is harnessed to an articulated affair probably built specifically to transport large pieces of timber, iron girders and sections of rail.

(National Railway Museum)

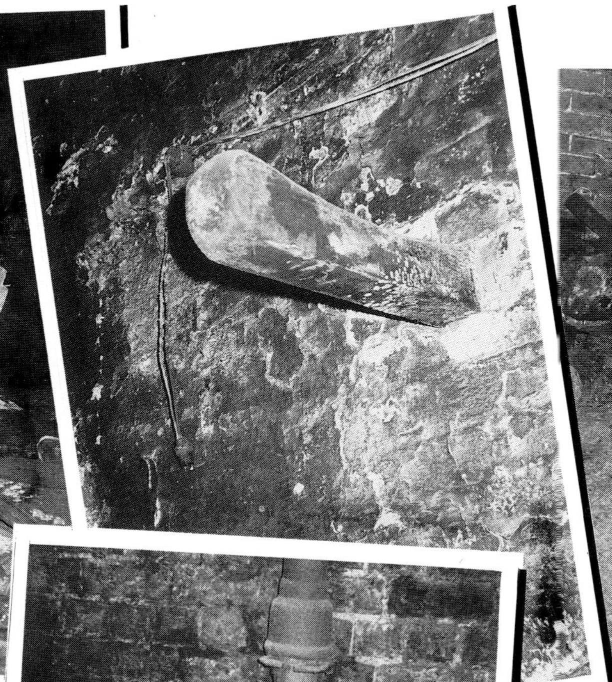

A few pieces of mouldering wood and rusting metal are all that remain of the once bustling shunt horse stables at the old Snow Hill station. One picture shows the wooden standings that divided the horses; another one of the wooden pegs on which the harnesses were hung. A remnant of a horse trough lies abandoned and a stove round which the men would warm themselves rusts quietly away against a wall.

(Bryan Holden)

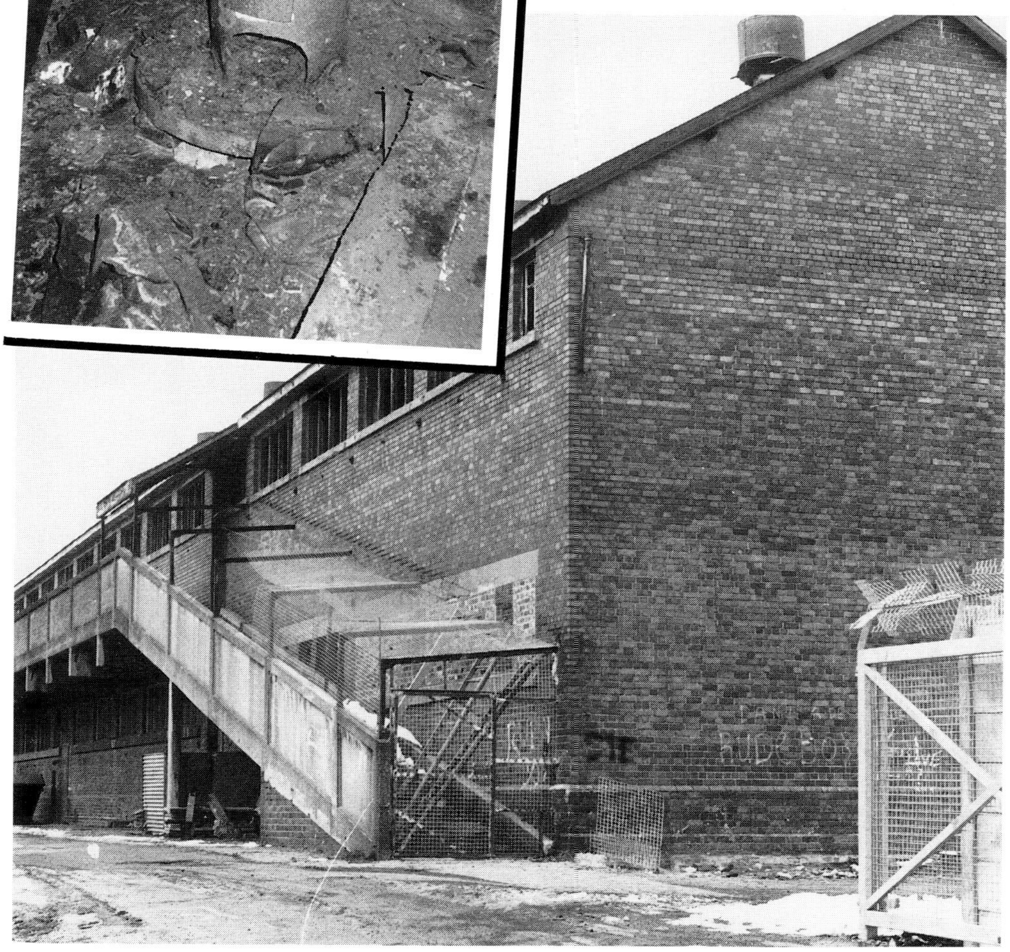

The Great Western Railway built these impressive stables at its goods yard at Hockley - and with 150 horses to accommodate, it solved the space problem by sending some of them "upstairs to bed" via this wooden ramp. The picture dates from 1981.

(Bryan Holden)

BIRMINGHAM'S

Along the 'cut'

WORKING HORSES

Canal horses at work were seen by relatively few people, rather they were glimpsed, fleetingly, from bridges as they plodded steadily along the towpath, a few yards ahead of a narrow boat cutting silently through the water at the end of a rope.

The horses were part and parcel of the secret life of the water gypsies, the canal boatmen and their families, who, like snails, took their homes with them wherever they went.

Boat horses were not the mighty heavy workhorses of the railway dray and the beer wagon. They were known as 'half-legged 'uns', a cross between a heavy and a light horse.

But a good boat-horse was an animal of intelligence and spirit. Old boatmen recall that some of the brighter sparks learned to take a breather by jumping into the canal and swimming, rather than hauling.

One of the best known boatmen of the Black Country is Alan 'Caggy' Stevens.

Lifford Canal Wharf is still there at Bredon Bridge on the Pershore Road, Cotteridge, but it has a derelict look about it nowadays. In the early 1900's it was all abustle with a constant flow of Midland Railway trucks linking up with the company's narrow boats.
(Birmingham Central Library)

In the 'fifties George Arnold and his son Jim worked the canal between Sandwell colliery and Ocker Hill power station. Their faithful horse Betty, seen here with them at the top of Eight Locks, Rider's Green would make the journey twice a day with a load of coal slack to feed the turbines. The photograph is one among many taken by Graham Guest - with a box-Brownie camera - when he was a lad living in Smethwick.

Middle & bottom right:
Two more pictures that capture the magic of the canals when narrow boats plied their trade and there was a serious job to be done. Both were taken at the top of Spon Lane Locks, Smethwick, in the summer of 1960.

(Graham Guest)

Below:
Caggy Stevens - Caggy because he is left-handed - is a legendary figure in Black Country canal lore. He loves the life and still works a butty boat on the waterways he knows better than most. He is a mine of information about the old days.

(Birmingham Post & Mail)

'Caggy' Stevens had a preference for former milk horses. 'Those that were too lively between the shafts of a milk cart made the best boat horses.'

The mileage of the canal system in and around Birmingham exceeds that of Venice. In its heyday the Birmingham Canal Navigation Company, linking the industrial Midlands between Wolverhampton and Birmingham, was a mainline water way.

Birmingham became the centre of a nationwide canal transportation system with many miles of almost continuous wharfage serving a meandering ribbon of factories and warehouses settled along the waterway.

The dawn of the railway era did not cut down the canal companies as it did the road coaching firms. Eventually, long distance canal traffic suffered, but short-haul work flourished and the railways built numerous rail/canal interchange basins.

But in the long run the railways got the upper hand.

Life on the waterways proceeded at a steady pace: it was not a case of rushing from place to place, dropping off one load, taking on another and being back home the same day.

Boats might be in such as Cambridge Crescent Wharf, Birmingham, for upwards of a fortnight unloading one cargo and taking on another. Formerly owned by the Shropshire Union Railways & Canal Company, latterly in the ownership of the LNWR, the Crescent Wharf provided stabling for canal horses and supplied them with corn when they left on long distance journeys.

In Gas Street, off Broad Street, there was a Day School for the boat children

Fred Dale and his wife worked a pair of horse-drawn canal boats for the Fellows Morton and Clayton company and were known as "jossers". Fred is seen driving into Giles Lock near Deakins Avenue, Witton.

Fred's son George, of Edgbaston, says: "We four lads had to act as mates for dad to bring two men's wages into the house."

The other picture shows a general view of Giles Lock.

An idyllic summer scene on a canal thought to be in the Birmingham area in the mid-thirties. A narrow boat is negotiating the lock while its one-horse power unit waits patiently on the tow path.

and a Sunday school, too!

The cargoes carried were of every description. Coal, limestone, tar, sugar, flour, soap and salt. The railway horses stabled at the Crescent Wharf (some sixteen in one stable with their names above each stall) would be required to deliver the goods which arrived by narrow-boat to local wholesalers and shopkeepers.

Cartloads of 2 tons of sugar were commonplace. The vehicles would return to the wharf back-loaded with heavy manufactured goods: scales from Avery's and ironwork of all kinds, for transhipment by narrow-boat to all parts of the country.

Large-scale workaday life on the canals came to an end in the early '50's. And with its demise an era of our industrial past - just a hazy blur in the memory of most of us - slipped unheeded into history.

But fortunately, private groups of canal enthusiasts and preservationists, together with enlightened local authorities are now increasingly aware of the priceless asset of our canals for recreational purposes.

As in other parts of the country, we may now look forward to seeing the modern-day successor to the old-time canal horse back on the towpath, bringing enjoyment to a new generation and re-creating a proud part of our industrial heritage.□

It was a red letter day for the children of Raddlebarn Road School on September 13 1934. For a shilling fee, they were given a trip on the nearby canal at Selly Oak. And there to draw the narrow boat, with harness jingling and brasses glinting in the sunshine, was, of course, a horse trained in the art.
(School Archives)

A rare sight in 1987... Chester pulls a butty boat alongside a canal in Birmingham city centre. The horse was used in a six-week trial cleaning out lock-gate rubbish and he passed with flying colours!
(Bryan Holden)

BIRMINGHAM'S

7 # Rolling out the barrels 7

WORKING HORSES

Powerful heavy horses were at one time an essential part of the brewer's trade. Often they worked in pairs, and their sheer bulk and decorous appearance was eye-catching.

The draymen always drew a circle of onlookers wherever they stopped to unload, and their skill manipulating coils of thick rope and handling lengths of bright chain was much admired.

The physical prowess of man and beast was a tangible thing. The patient

The draymen pose in their hessian aprons for cameraman Alfred Juggins to capture for all time this magnificent photograph as new supplies of Mitchells & Butlers beer arrive at the New Inns. Roger Davis, of Wylde Green, who owns the picture, knows that his grandfather, George Davis, is one of the draymen standing there. The picture was taken around 1904.

horses putting their great strength into drawing the heavy load. The corduroy-trousered draymen, long hessian aprons bound around ample waists, lowering enormous iron-hooped barrels into deep cellars.

For most people, the massive Shire horse was synonymous with the brewery dray. But Mitchells and Butlers, the famous firm of brewers at Cape Hill, Smethwick, preferred the more compact Suffolk Punch.

In 1946, there were 32 heavy horses working at Cape Hill. Six of these were Shires and the rest Suffolks.

The Suffolk horse is an excellent mover with a smart, quick step. But it was the rule that M & B's horses never trotted. They were well able to stand a long day's work, but it was rare, indeed, for any journey to exceed one mile from the brewery.

The blacksmith's shop at the brewery was above the stables, a somewhat unusual location, where all shoes were made and fitted by the firm's own blacksmiths.

There were also shops for joiners, saddlers and wheelwrights. A colourful sight often seen through the street railings was an array of brightly painted cartwheels, of varying sizes, set in a long line against the brewery walls, ready for re-fitting.

Brewery horses were familiar through their ceremonial appearances. In the 1939 Birmingham Horse Parade ten of M & B's Suffolks took part, together with a pair of high-stepping Hackneys, pulling a spirit delivery coach.

Their rivals, Ansells Brewery, of Aston Cross, were represented by four Shires: Bonnie, Bob, Roger and Major, names familiar to many.

Unfortunately, although several London breweries have demonstrated that the horse is still an efficient and economical means of short-haul town delivery, there seems little likelihood of a return to horse drawn transport in Birmingham. □

Davenports began its Beer at Home service in 1904. The crates of bottled bitter became a familiar sight for Birmingham people for many years until supermarkets ruined the trade.

The brewery itself began in 1829 and remained independent until 1986 when it was taken over by Greenall Whitley. Here's a line-up of Davenports delivery vehicles about to set off.

(Birmingham Post & Mail)

Asked to guess the date would you win a prize? Probably not; despite the Edwardian look of the Mitchells & Butlers horse-drawn dray, the glimpse of a car on the right gives the best clue. The picture was taken in Birmingham's Paradise Street on July 11, 1938.
(Birmingham Central Library)

For 25 years Fred King has been under contract to shift rubbish for Ansells Brewery. Now, in August 1960, he is the last of the horse tippers and he is about to be pensioned off - so he poses for a photograph with his faithful horse Dolly. Mr N Fisher of Great Barr supplied the picture.

MITCHELLS & BUTLERS
BREWERS

Wagons roll! The Mitchells & Butlers' beer wagons are reminiscent of a Wild West film . . . This superb picture shows all the hustle and bustle at the Cape Hill brewery, Smethwick, as the drays pick up their loads for delivery to Birmingham's countless pubs. Date unknown.

(Birmingham Post & Mail)

Holding the reins is Bob Hawkins, a Mitchells & Butlers delivery man, employed by the brewery from 1934 until he retired. With this turnout, says his daughter, Mrs Jean Woodfield, of Oldbury, Mr Hawkins won the Durley Trophy for the best entry on rubber tyres.

The shape of things to come is evident in this photograph taken at Ansells brewery at Aston Cross . . . motorised vehicles are begining to outnumber the traditional horse-drawn drays.
(Birmingham Post & Mail)

BIRMINGHAM'S

The law in the saddle

WORKING HORSES

The good thing about police horses is that they have a past, a present, and a future, too. And looking at old photographs, and digging into archival records, we can learn something of their illustrious history.

But, best of all, every day one can see police horses going about their job of work, and feel happy in knowing that they are not a threatened species in the shadow of extinction.

The headquarters and stables of the West Midlands Police Mounted Division are in Park Lane, Aston, Birmingham. The division took-up this purpose-built accommodation when the old Duke Street stables were vacated (and almost immediately demolished) in January 1984.

Chief Inspector James Reed leads the Birmingham 'greys' in their march past HM Inspector of Constabulary in 1953. Mr Reed is happily still active at 92.

The famous greys of Birmingham's mounted police line up with their riders in 1937 for the annual inspection by the Lord Mayor, then Alderman Harold Roberts. With him on the right is Alderman H P Sayer and in the centre Major Brennan de Vine, chief of the city's Veterinary Department. The major played a key role in the continuing success of the Birmingham Horse Parade.

Over the last four years, the establishment has grown from 29 to 40 horses. Twenty are stabled in Park Street; nine at the 'Tally-Ho' Police Training School, Pershore Road; and a further 11 horses at West Bromwich.

Surprisingly, it was not until 1926, in a time of nationwide industrial disorder and social unrest, that the Birmingham City Police was granted its own horse establishment, with a trained and regular complement of officers.

Before that time, mounted duty had been a somewhat haphazard and ad-hoc affair, carried out by foot-duty policemen with mounted experience gained in cavalry regiments of the British Army. It was the normal practice to borrow horses for police work from local undertakers.

In 1850 the new Duke Street stables were completed,but not for another 75 years did they become the permanent home of the City's police horses. Prior to 1926 they housed only those horses which drew fire engines, ambulances and prison vans.

With the establishment of the Birmingham City Police Mounted Branch, it

The West Midlands mounted police have always been noted for their 'greys' but, in 1958, as the number of breeders falls, it becomes more difficult to specify the colour you want. Hence, a dark horse has literally got in on the act in this line up at the traditional visit of HM Inspector of Constabulary.

Note the sleeve badges which denote that the wearer is a qualified St John Ambulance first-aider. In times of crisis the cry was often heard 'Get a copper and make sure he's got a wheel on his arm!'

Birmingham's mounted police horses were housed for several years in stables at Duke Street, Aston, and here's a smart line-up outside the premises. The building was erected originally for ambulances, prison vans and fire brigade horses. Just visible is one of the horse troughs that lined the outside walls. Further along a window has been roughly boarded up. It was a sign of the times - a few days later Duke Street was closed down.

Mounted policewomen are a modern extension of this method of keeping law and order. This picture was taken by the Birmingham Post & Mail in November 1988, when Policewoman Jo Dixon of the West Midlands force, received the cup as best horse and rider.
(Birmingham Post & Mail)

Bought in 1958 as a four-year-old, Grey Sky (above) matured into a champion horse. It won every major police trophy in the country as well as the Wembley Horse of the Year award for best-trained police horse. Ridden here by Pc Picken, Grey Sky was retired in 1979, aged 24.

became policy that only greys would be recruited, mostly cross-bred Irish draught horses. Collectively they were known as the 'Birmingham Greys', or simply, 'The Greys', and became famous throughout the country.

After the war, with the general run-down in horse breeding, it became difficult to obtain enough grey horses of the right conformation. Horses of other colourations, but usually bay, entered the ranks.

But grey horses were still much sought-after. And one particular grey horse, which came from Ireland as a 4-year old gelding in 1958, was far ahead of its contemporaries. It was called 'Grey Sky', and went on to become a champion. Grey Sky won every trophy that was going, crowning his achievements with the coveted title of Best-trained Police Horse, beating all-comers at the Horse of the Year Show, Wembley.

Currently, there are a number of rising stars in the Park Street stables, including 'Memsahib' a bay mare, winner of the best-trained novice award at Cleveland in 1987. Waiting in the wings is 'Paragon', a massive cross-bred hunter (18 hands) saved from the knacker's yard. Paragon has the neck and chest of a Shire, and is being considered as a future drum-horse.

Chief Inspector Des Turner, head of the West Midlands Police Mounted

A new event at the 1989 Birmingham Horse Parade is the Musical Ride by the West Midlands Police Mounted Branch. At the centre is the grey drum horse Javelin, seen here ridden by Pc Ingham. Javelin was bought as a five-year-old from a farmer in Gloucestershire and has been with the division since 1979. In the Musical Ride, mounted police show incredible skill and precision as they turn, circle, pivot, and cross. Full dress uniform and red and blue pennons add to the colourful display which reaches a climax with a charge and salute.

Division, stresses that police horses are always working tools. 'Mounted policemen do every policing job, same as the foot-constables, except that they work on horseback,' he says.

'In the public mind the police horse is seen controlling football crowds and outbreaks of civil disorder. Certainly there's nothing to beat a trained horse and rider for public order work.

'But it's not all confrontation. The horse is ideal for controlling large crowds; dividing them into manageable groups; helping to avoid the build-up of tension.'

There's no doubt that the presence of mounted police at football matches and protest meetings is a powerful psychological influence in keeping order. But the long white batons are part of the image. They are never used to crack heads.

So, why does a mob rarely attack the horse's legs in order to unseat its rider?

'The prime target has always been the man' said Chief Inspector Turner. 'It was so in mediaeval times, when the horse wore armoured plate to guard its head and chest, but its legs remained unprotected.

'The assailant always goes for the man. But the horse's head is vulnerable, which is why we equip our animals with nose guards and eye shields.'

One of the most popular police displays is the Musical Ride, performed by

This Evening Mail photograph graphically captures the most infamous incident ever to shame a football pitch in Birmingham. It happened on the St Andrews ground on May 11, 1985, when fans rioted on the pitch during a Blues v. Leeds match. Birmingham's mounted police were called in to restore order and in scenes reminiscent of a cavalry charge saved a very serious situation from devloping into a major disaster. Once again the mounted police had proved their worth in safe but emphatic crowd control. Even so, a 15-year-old boy was killed, when a wall collapsed, and 88 fans and 148 police officers were injured in the riot.

It's an odd fact that hooligans bent on attacking mounted police go for the policeman - or the horse's head. With this psychological quirk in mind, the horses are given eye and nose shields - and the men wear crash helmets. As in the pitched battles of history, the steeds' legs go unprotected.

An advantage of the police horse is that it can go where motor vehicles cannot. Here a mounted patrol surveys scrubland - and the riders get a first class view from their elevated position.

16 horses of the West Midlands Police Mounted Division.

The horses and riders demonstrate turns, circles, and a repertoire of pivots, passagings, sunflowers, and the spectacular Maltese Cross. Star of the show is the grey drum horse 'Javelin', with the division since 1979. He was bought as a 5-year old from a farm in Gloucestershire.

The West Midlands Police Mounted Division is a self-contained community. One senses an air of quiet efficiency. Only in this ordered and dedicated manner can the division function as a compact and supportive unit, able to handle one of the most arduous and, at times, most personally hazardous tasks in all police work. □

BIRMINGHAM'S WORKING HORSES

Making muck into money

Before the introduction of the water closet, the night soil overflowing from the privvies, cess-pits and middens in the fast growing town of Birmingham was a serious health hazard.

Sewage disposal became a throbbing headache for the hard-pressed Public Works Committee until, in the 1870s, with the laying of sewers and street drains, the Tame and Rea District Drainage Board came about.

In 1878, the Board started a sewage farm on 260-acres at Tyburn. It was to assist in the work of purification and treatment of sewage, not to make a profit out of farming activities.

Nevertheless, the acreage under cultivation grew rapidly, nurtured by the very sewage that it sought to dispose of. By 1910 the farm covered 2000 acres, and farming operations were at their maximum.

Bumper crops of hay, rye grass and mangolds were harvested; also good

Cutting the corn at Park Hall Farm, Minworth, where the Drainage Board carried out its farming operations after being forcibly moved out of Tyburn in 1918. (Severn Trent Water)

Not water water everywhere but horses! This was the daily scene at the stables at Tyburn where the Birmingham Tame and Rea District Drainage Board kept its animals. That was until 1918 when the War Office took possession. The Board's involvement in farming became so extensive that at times it seemed to overtake its primary responsibilities of drainage and water management.

In the years 1901-3 the Board farmed over 2,000 acres - and was losing £3,000 a year on the deal!

(Severn Trent Water)

Right: Harvesting at Park Hall Farm.

yields of mixed cereals and roots - even rhubarb. In one year the milk yield from Tyburn's 200 cows was 156,000 gallons. Thirteen of the animals gave over 1,000 gallons each.

Horses were involved in the agricultural side of the Board's work, which included ploughing, crop collecting, transporting materials, and the collection of sludge from the drying beds.

Farming at Tyburn continued successfully until the outbreak of war. In 1914 the War Office took over the buildings, as well as part of the farmland, and the Board had to sell off most of its stock and farm implements. With the sale went not only the horses, but workshops for joiners, blacksmiths and wheelrights.

The Tyburn premises were finally vacated in 1919 and let to the Dunlop Rubber Company. The Board's agricultural work was transferred in a reduced form (161 acres) to Park Hall Farm, Minworth. Farming operations there came to an end in 1923.

The Tame and Rea Board continued to use horses for sludge disposal, and as far on as 1939 four of its equine workforce, two mares and two geldings, turned out with their tipping carts in the Birmingham Horse Parade.☐

May 1, 1934, and two of the Drainage Board's working horses at Minworth are decked out in shining harness and gleaming brass for the May Day Parade.

BIRMINGHAM'S WORKING HORSES

Caring about horses

Before the 1939-45 war there were some fifteen RSPCA clinics in the Birmingham area with qualified veterinary surgeons undertaking 'free or assisted treatment of animals of the poor'.

The Birmingham and District Branch took a keen interest in the welfare of horses, and its inspectors kept a daily vigil on the streets for animals in distress.

Back street stables, where horses suffered from bad food, poor stabling and inefficient grooming, were a particular target, and the Society's inspectors

On July 24, 1922, the RSPCA opened a Home of Rest for retired horses in the green fields of California, at that time outside the Birmingham boundary. Here are some of the "pensioners" grazing happily in the sunshine. (RSPCA)

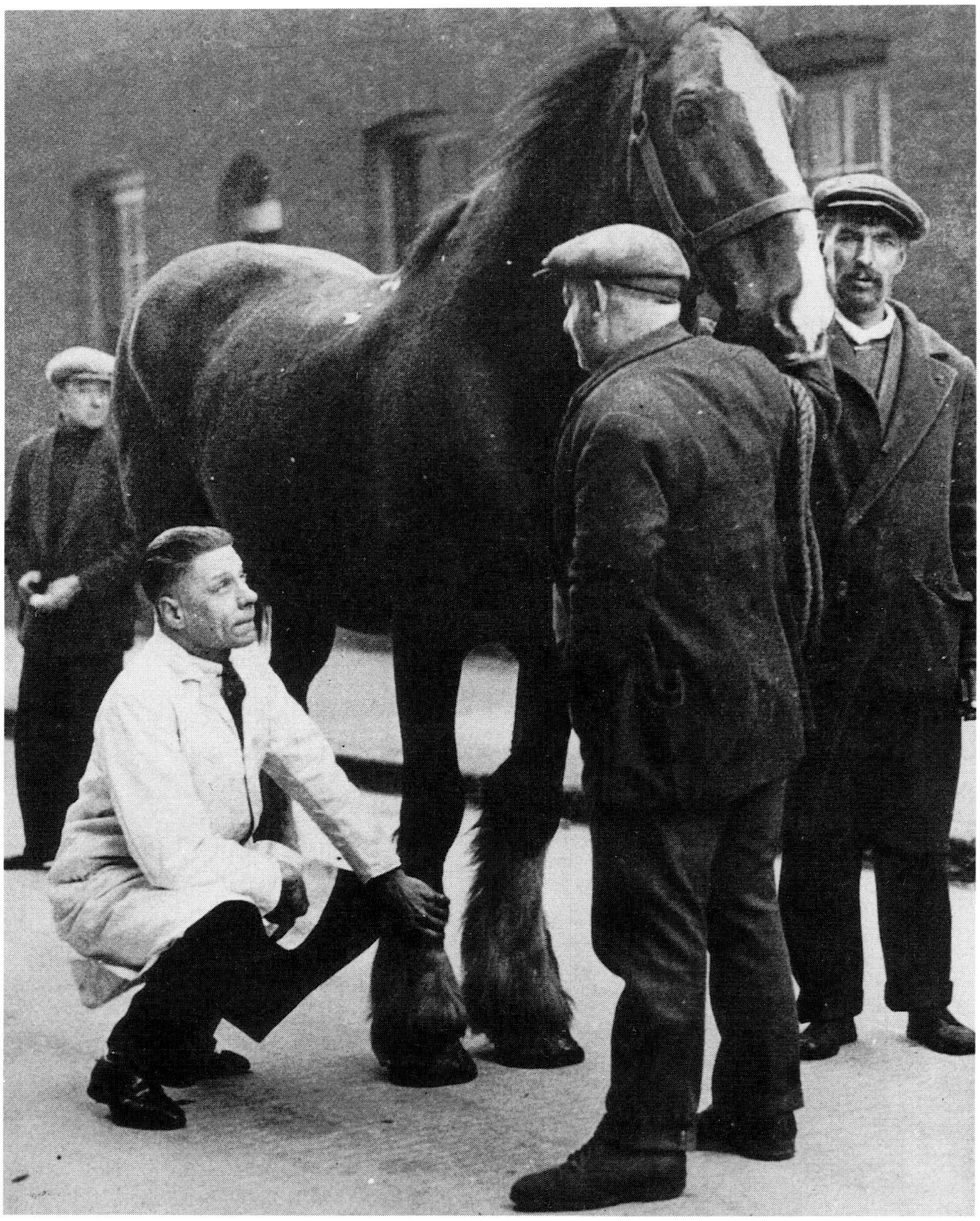

Inspectors from the RSPCA were always on the look out for horses that showed signs of ill-treatment and would carry out checks in the street. Their ministrations were taken seriously, as is evident from the tension in this photograph.
(RSPCA)

ventured forth at the risk of confrontation with incalcitrant owners.

The RSPCA was a prime-mover in the organisation of many shows and parades and, to stimulate the interest of those in charge of horses, presented merit badges and certificates. These badges were greatly prized and fixed to the harness for all to see.

Commenting on the Birmingham Horse Parade, in its Report and Accounts 1937, the RSPCA stated: 'We are anxious to draw attention to this very fine parade, with the object of fostering greater interest in the well-being of the very large number of horses which are daily working in the City. Such parades, undoubtedly, do a great deal to encourage drivers and stablemen to show kindness to those noble creatures entrusted to their care.'

Also in the report was a plea to motorists to give-way to horses in crowded streets and on hills. 'We fear sometimes that motorists do not realise that horses have no gearbox or brakes, and that stopping and starting depends upon their muscles. Often the road surface is glass-like, and yet they must obey the traffic lights'.

Both the RSPCA and the City Veterinary Department kept a close watch on abbattoirs and slaughterhouses. Throughout the 1950s hundreds of Birmingham horses, many in their prime, ousted by the motor vehicles, were turned into hides, bone meal and dog food.

Prices paid for live animals reflected the glutted state of the horseflesh market. A receipt dated 2 July 1951, for three bay geldings and a chestnut mare (alive) totalled only £54. This compared with a price of between £30-£40 paid by the Public Works Committee in 1859 when it was buying a new horse.

To encourage the use of undocked horses, and to discourage the brutality of tail-docking, the Society presented a Silver Challenge Cup. It was hoped this would be a means of enlisting public support for the Bill then about to come before the House of Commons.

In 1937, the cup was won by Mitchells & Butlers Ltd.☐

The infamous Hame or Bearing Rein prevented the horse from dropping its head. Wearing this totally unnecessary leather strap robbed the horse of the use of its weight in helping to pull its load and caused avoidable distress. The RSPCA led the fight to have the hame rein abolished, as can be seen from the cutting and poster reproduced below.

Another RSPCA campaign was carried out against the practice of docking horses' tails. To encourage owners to stop it, in the mid-thirties the society provided this cup to be competed for at the Birmingham Horse Parade. It could be won only by undocked horses and the RSPCA gave it "in the sincere hope that it will not only draw attention to this matter, so vital for the animals themselves, but will also be a means of enlisting public support for the Bill which it is hoped may soon come before the House of Commons."
(RSPCA)

The Royal Society for the Prevention of Cruelty to Animals pursued a long and vigorous campaign to stop the maltreatment of horses.
These posters were typical of many issued by the society in the mid-thirties.

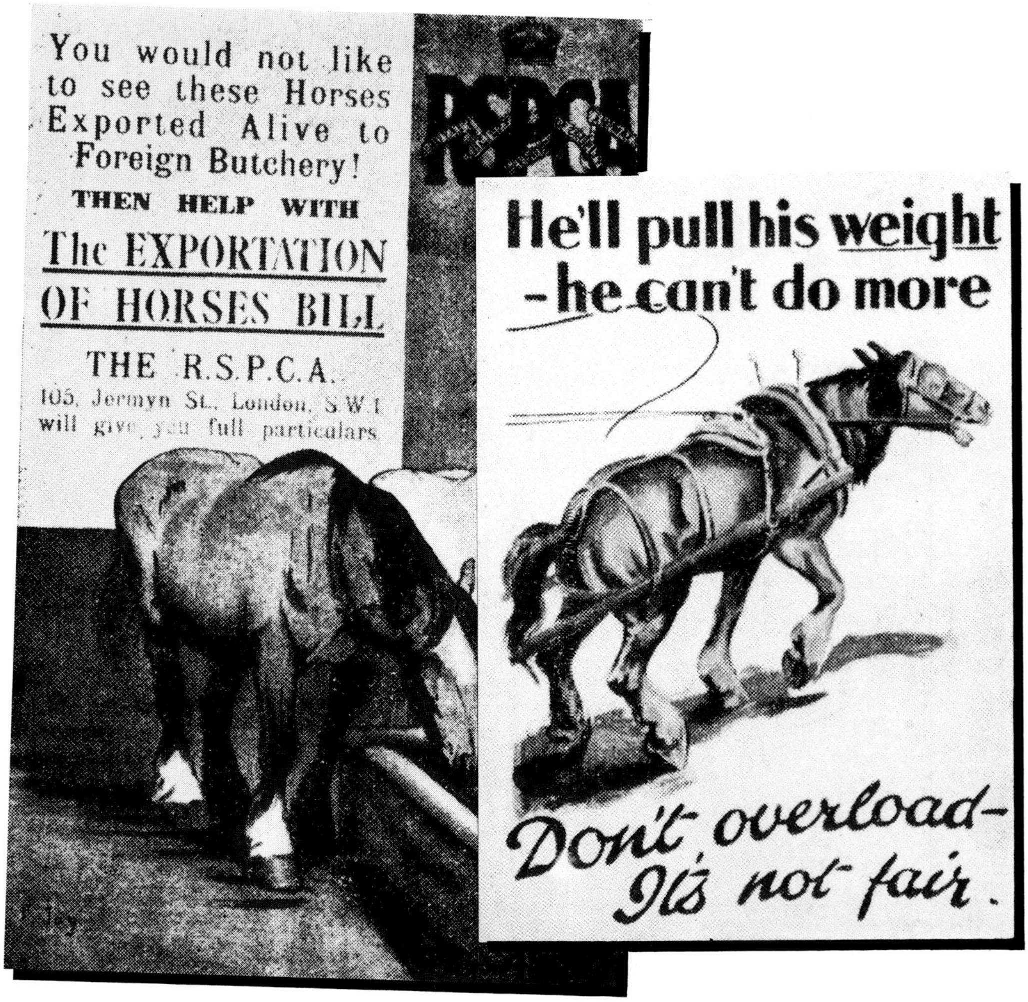

The RSPCA was successful in helping to ban the docking of horses' tails. The docking knife (1) was the cruel implement used by owners more concerned with their horse's looks than its welfare.
(Bryan Holden Collection)

Bleeding was at one time considered a health aid in horses as well as humans.
The fleam hammer, or meld (2) and the fleam knife (3) were the implements used to carry out the operation on horses. A pernicious act based on ignorance.
(Bryan Holden Collection)

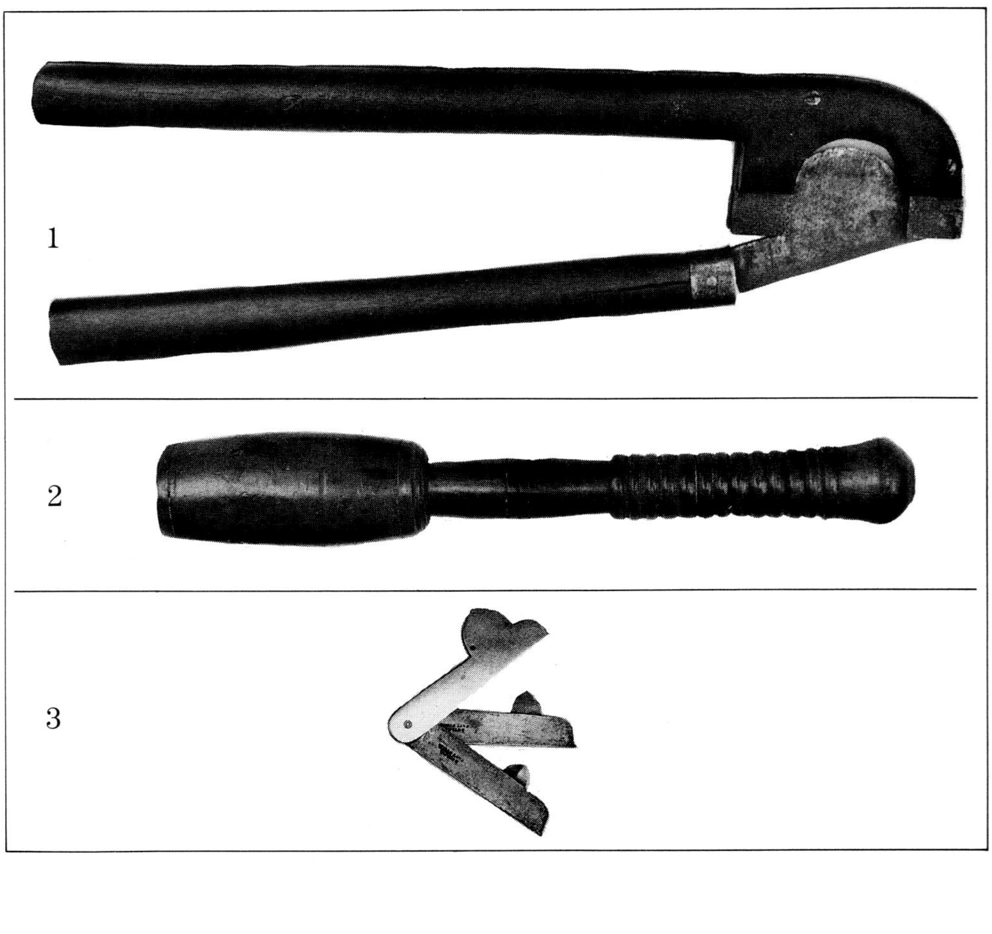

Two more photographs of the RSPCA Rest Home for Horses at California. They show that the society catered for animals of all sizes and breeds.
(RSPCA)

Kate, an LMS parcels van horse, proudly wears her merit badge in the 1933 LMS Horse Parade.

The photograph was taken in Queens Drive, New Street Station. Kate was then 19 years old and well-known in central Birmingham.

By chance we also came across this newspaper cutting of Kate in retirement at the RSPCA home, California, Birmingham.

Looking on is Vanity, a war veteran. The photograph appeared in the Sunday Mercury in December, 1935.

BIRMINGHAM'S WORKING HORSES

11

The proudest of them all

Aristocrats of the horse world, before the 'universalisation' of the motor vehicle, were the proud, rather arrogant-looking black horses employed by undertakers.

One of the best-known firms in the south of Birmingham was William H Painter Ltd., which began in a small way in the early 1900s. In those days, in addition to funerals, Painter's ran a tuppenny (½p) horse brake between The Swan, Yardley, and The Wheatsheaf, at Sheldon.

Perfection in every detail . . Mr Charles Rogers, head horseman for funeral directors William Painter, holds the reins of two elegant Belgian black horses.
(Mrs Dolby Collection)

Outside the Painter's funeral business headquarters, then at the corner of Church Road and Coventry Road, South Yardley, in about 1910.

The signboard outside the firm's premises, then on the corner of Church Road and Coventry Road, South Yardley, advertised 'Hansoms, Landaus and Charabancs'. The funeral horses also turned out for weddings, often on a Bank Holiday or on Boxing Day.

Chairman Mr Bernard Painter, whose grandfather founded the firm, recalls boyhood days helping in the stableyard.

'New horses were broken-in on Sunday mornings, run around at the end of a rope. Lungeing, it was called.

'Sometimes the horses were quite spirited, after having been brought straight in from the Chipping Camden sales'.

In the 1930s the firm moved to new premises on the opposite corner, and at its zenith the stables there held 20 Belgian black horses, with sweeping long tails, profuse manes and proudly arched crests.

There was a staff of twelve coachmen.

The stableyard at Painter's was a self-contained world with its own harness room and smithy. The forge and the giant bellows are still there, but the one-time blacksmith's shop is now a storeroom.

In frosty weather, drivers came in early in the mornings to put frost cogs into their horses' shoes to stop them slipping. And every night they cleaned their own harness and hung up the collars so that the straw filling would be sweet and dry for work next morning.

'It was a beautiful sight', recalls Mr Painter. 'Rows of shining black leather harness gleaming with bright fittings and polished buckles'.

Painter's horses took part in many shows and parades, particularly the annual carnival procession through Solihull. William Painter ('The Boss') was a keen participant himself, always seen walking alongside his turnout, a rolled black umbrella in his hand.

The firm finished with horses in 1938, but hung on to a pair and a coach for a few months more. But the motor car had finally come into its own and it was the done thing to have a motorised funeral.

In the old days an impressive funeral was not for the rich alone. A team of black horses, an eye-catching carriage with a glass-sided hearse to show off the floral tributes, and a convoy of mourners following in closed carriages, was an essential part of a socially acceptable departure. □

The black horses are there, and the Painter's black coaches - but so are the flags and bunting, proving that undertakers were not averse to a little levity when Solihull carnival time came round in 1925! And walking at a brisk pace in front is Mr William Painter. The picture comes from Mrs M H Dolby of South Yardley, whose father, Mr Charles Rogers, is at the reins. He was the firm's first employee and he stayed all the 40 years of his working life.

(Mrs Dolby Collection)

A carriage for every occasion was the policy of Painter's. Though essentially in the funeral business, they also provided suitable transport for weddings or any other big social event and always in the very best of taste. This was one of their 'landauettes,' seen here in Church Road, South Yardley, drawn by two elegant black horses,

(Wm H. Painter Ltd)

Dignified perfection is expressed by this fine black Hansom cab drawn by a horse whose white socks and blaze were in dramatic contrast to its black coat.
(Wm H. Painter Ltd)

Jet black horses, a stately black funeral carriage, and a driver in a top hat...
(Mrs Dolby Collection)

BIRMINGHAM'S

12

A century of spectacle

WORKING HORSES

The origins of Birmingham Horse Parade go back to the early 1890s. But the date of the first parade remains unknown, despite hours of trawling through old newspapers and public records.

The run-down in horsed transport began to show itself particularly after the 1914-18 war, but the importance of the Birmingham Horse Parade was such that on May Day 1938 it still attracted nearly 250 entrants.

At the close of the judging there was a parade through the streets and more

For months these sturdy workhorses, sweatstained and begrimed, have hauled Co-op coal through the city streets. But now it is the big day for them and for their masters. Today they are pulling the world no less - well, a model of it - in the great Birmingham May Day Parade of 1934. Standing proudly at the helm is Hubert Stephens, who received many a medal for his artistry.

Picture supplied by son-in-law Dennis Goldie of Fordbridge, Birmingham.

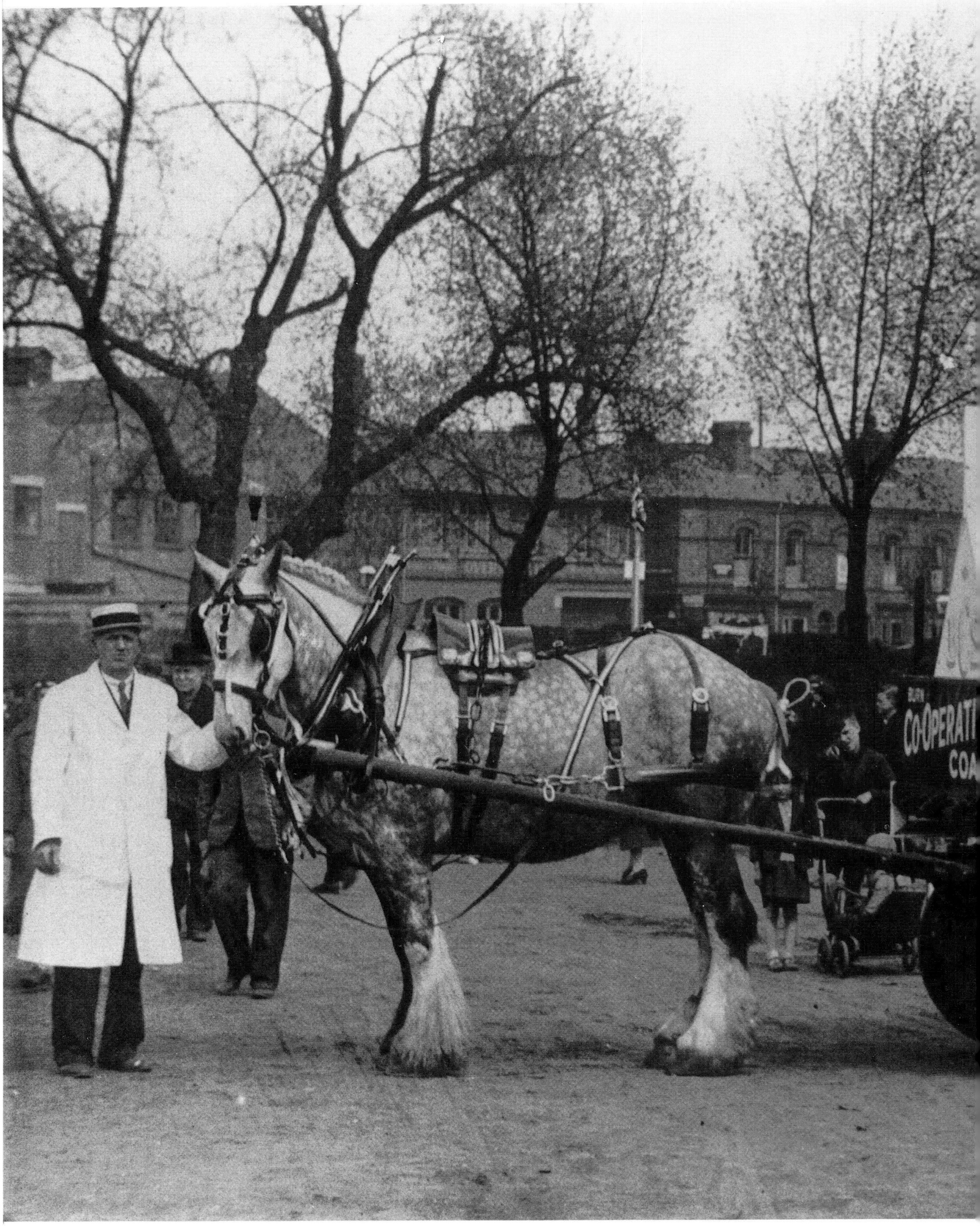

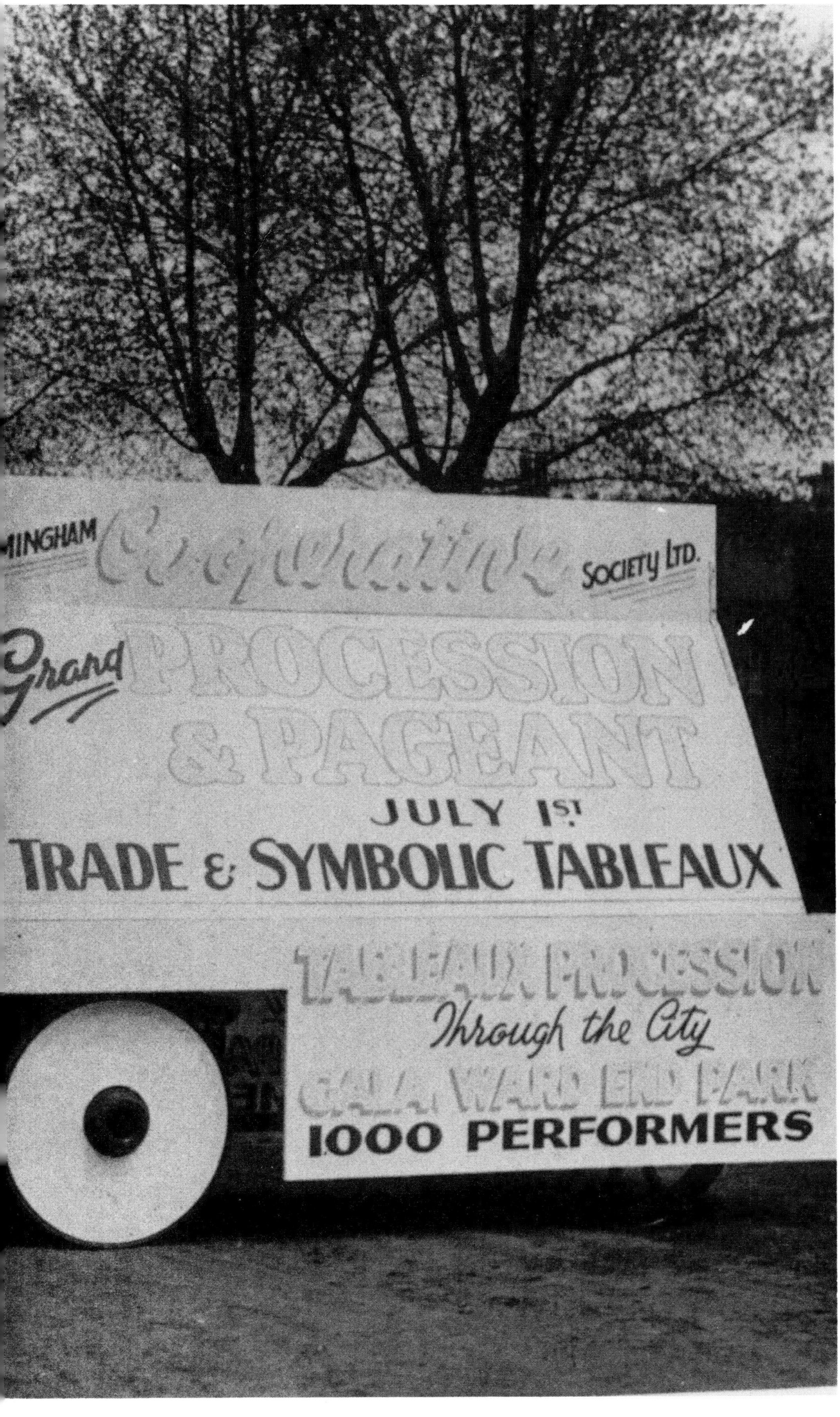

When Birmingham Co-op held a Grand Procession and Pageant in Ward End Park, it advertised the event with this adapted coal wagon pulled by a dapple grey. How many of the "1,000 performers" were horses we do not know!

The photograph is from a collection compiled by Jim Cowell when he was the Co-op vet, and like two others elsewhere in this book, it was provided by Mr Cowell's daughter, Miss Rose Cowell, of Handsworth.

than a mile separated the mounted police leading the procession and the last vehicle.

Sir Frederick Smith, president of the organising body, gave a luncheon at the Midland Hotel to supporters of the 1938 parade. He spoke of the high quality of the horses:

'We have succeeded in getting owners - big companies and others - to help us to show horses, properly fed, well-groomed and cared for - horses, in fact, of which the city can be proud.'

The parade assembled in Calthorpe Park, Balsall Heath, and proceeded down Pershore Road, Belgrave Road and along Bristol Street.

After moving through the principal streets packed with sight-seers, and receiving civic acclaim from the steps of the Council House, the parade went along Paradise Street and into Broad Street. It dispersed in King Alfred's Place, now the site of the Birmingham International Conference Centre.

But the Big Parade wasn't just for the top names. There was a special class for family butchers, greengrocers, ironmongers, and the smaller shopkeepers.

Mounted stewards were provided by the Legion of Frontiersmen, and some

Mr H C Dye of Olton, was a saddler and harness maker in Birmingham from the time he started as a half-a-crown a week apprentice in 1914 until he retired in 1970. These two photographs illustrate some of the kinds of leather craftsmanship he produced. The four-wheel dray belonged to Mitchells & Butlers. The grey Shire, says Mr Dye, was owned by Arthur Lewis, "one of the best breeders and judges of this particular breed".

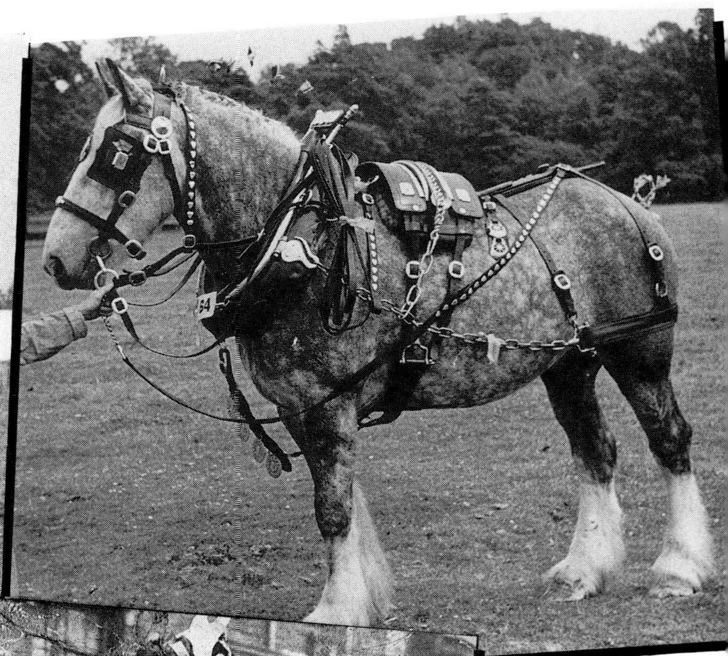

It's May 1, 1926, and young Tom Wolsey is all set to take his beautifully dressed horse from the family coal business in Clark Street, Ladywood, to the parade in Calthorpe Park. Now 79, Tom says he used to deliver coal for his father Jack, and in his spare time chance his arm as an amateur boxer.

twenty officers and troopers competed for a challenge trophy in a class of their own.

The legionnaires were a colourful sight in their dark blue tunics and fawn breeches. Their founder, Captain Pocock, had served as a scout in the Boer War, and had adopted the distinctive Baden-Powell bush hat.

They showed the old flag at gymkhanas, shows and fetes, bringing a whiff of past adventure to modern times.

Away from the big annual parade, many private shows and stable

Mr F Poole of Hollywood, near Birmingham, joined the Legion of Frontiersmen in the late 'forties. He supplies this photograph of himself (fourth from left) and colleagues on parade in Selly Oak Park.

Bert Oldham was a sergeant in the Birmingham branch of the Legion of Frontiersmen and here he is at a horse parade in the city just before the last war. His nephew, Howard Bayliss, of Kings Heath, recalls seeing him and other legion members playing in a band.

PRICE 3D.

BIRMINGHAM HORSE PARADE

PRICE 3D.

SATURDAY, 13th MAY 1939

OFFICIAL

PROGRAMME

The General Public Parade of Horses and Vehicles through the Principal Streets of Birmingham, is supported by Horse-owning Members Comprising Co-operative Societies, Bi mingham Corporation, Brewery Companies, Milk Traders, Bakers and various other trades.

Judging of Exhibits will take place at Calthorpe Park Extension at 3-0 p.m.

Hon. General Secretary: BRENNAN DE VINE,
Telephone: MIDland 0121. 7 St. Martins Lane, B'ham, 5

competitions were organised by the railways, and the larger milk, bread and coal retailers, to give every horse and driver in the stable a chance to win a prize. In effect, these were competitions to sift the best turnouts for the big parade.

Pride and a close friendship developed when man and horse spent every working day together... Here's Henry George Twigg, roundsman with Fowler's Forest Dairies at Sparkhill, at a show in the 'twenties. His son Leonard, of Sheldon, provided the photograph.

Now it's 1931 and, (below) the Birmingham Co-op holds a parade, this time to mark its Golden Jubilee. In the procession was this remarkable entry which displayed the "Cannock and Rugeley Nuggett," the largest piece of coal to be raised intact from any pit in England. Its weight is not known now, but it took three horses to pull it! And the onlooker who guessed its weight correctly won it!
(Central Midland Co-operative Society Ltd)

Left: Arthur Lewis's Shire horses, bred at his stables at Tanworth-in-Arden, have become renowned throughout the world. Here he is at the Three Counties Show at Worcester in 1952 with a pair of brown geldings that won the Heavy Turnout prize.

Below: Sid Rawlins, one of Birmingham's leading hauliers, and his head horseman, Jim Rumbold, proved to be an all-conquering team at Midland shows. They are seen with the trophies won in one season by just one of their horses, Bromford Supreme. They are together again in the other picture, on their way to another triumph with a double entry in a show at Harborne.

For 43 years Ted Jones worked for King Bros, coal merchants, of Brewery Street, Aston.

Ted loved nothing better than to spruce his charges up for the annual May Day parade - and there he is with grand-daughter Yvonne after one such outing. Photograph supplied by Mrs D Greathead of Kingstanding.

Jim Cowell was the Co-op vet when the company kept about 700 working horses in Birmingham. As such he gained a superb collection of photographs along the way and here are two of them of outstanding Co-op wagons. They come from Mr Cowell's daughter, Miss Rose Cowell, of Handsworth.

The pride of Scribbans' stable line up at the Birmingham Horse Parade in the early 1950's. These were the entries all the others had to beat - left to right - Verity, Odelle, Muriel, and Regent. They spent the more humdrum part of their working lives pulling the distinctive Scribbans' vans to deliver bread in Birmingham suburbs. The picture comes from Mr S Millard of Ladywood, who was with Scribbans for 27 years.

By the early 'fifties the end of the horse era was in sight. The City of Birmingham Horse Show, at Handsworth Park, lingered on with only a handful of commercial entrants. In September 1954, there were only two classes: railway horses (heavy goods) and single horses (light trades).

It was a peculiar combination of circumstances, not least a world war, and the comparative costs of fodder, fuel and labour, that had kept the horse and cart on the job for so long. □

Not all delivery horses were show-stoppers, but their masters had fun anyway. Here's a line-up of "also rans" from Handsworth Dairy at Selly Oak carnival in about 1948. They're all dressed up in crepe paper with the desperate hope it would not rain and spoil it all. If you won you were likely to get a rosette and 20 Players! The picture comes from former roundsman Bert Ward of Northfield.

Mrs Grace Pullinger, of Castle Bromwich, was an office worker for the Co-op Dairy in Vauxhall Road. She never went to a big parade but made a passing acquaintance with the horses...

On May 5, 1951, the Birmingham Co-op had a triple celebration - its 70th anniversary, May Day, and the national Festival of Britain. And of course, the Co-op horses were on parade again. Here are two entrants.

The Birmingham Post AND JOURNAL

MONDAY, MAY 16, 1938

BIRMINGHAM HORSE PARADE

A MILE-LONG PROCESSION THROUGH THE STREETS

MISS GWEN LALLY AND THE CENTENARY PAGEANT

No doubt it is true that in the end horse power will be displaced by fuel power, but things do not happen so quickly. It took steam a long time to kill sail. Nearly a century ago it was still a matter of heated argument whether sail or steam was the faster and better way o crossing the Atlantic; but even then the en of the controversy was sure.

From the Birmingham streets in an age o mechanical transport horses have not complete disappeared. To country-bred folk walking alo the pavements they are objects of admiratio and when they eventually vanish from the urba scene most people will feel that they ha suffered the loss of something that attach them to more spacious and romantic days. T extent of that loss could have been apprais by those who attended the annual Birmingh horse parade on Saturday at Calthorpe Park

The entries numbered nearly two hundred fifty—animals of cleanly and swift moveme with sleek skins that glistened in the sun, vehicles behind that were bright in their c of new paint. Everything was burnishe blackened to the last degree, and there w delightful jingling of pendant brasses. whips were festooned with rosettes, the to of past victories.

Birmingham Pageant

There were all sorts of horses and they of all colours. Present were those who t bakers, butchers, coal merchants, brewers, sellers, railway companies and for the Birmingham Corporation. They were displayed amid ideal spring conditions, and seemed more at home in the park than in following the daily routine in the congested streets of the city. At the close of the judging there was a parade through the streets, and more than a mile separated the mounted police heading the pro-

The writer of this report from The Birmingham Post of Monday, May 16, 1938, began his description of the big Horse Parade in prophetic vein... "No doubt it is true that in the end horse power will be displaced by fuel power..."

The parade itself was undoubtedly a magnificent spectacle as nearly 200 horses formed a mile-long procession that wound its way from Calthorpe Park to the city centre.

A VIP guest on the day was Miss Gwen Lally who was then busily planning a gigantic pageant to take place later in the year in the grounds of Aston Manor to mark Birmingham's other centenary - the granting of its Municipal Charter in 1838.

BIRMINGHAM'S

How the Big Parade came back

13 13

WORKING HORSES

by Peter Ricketts

The Big Horse Parade returned to Birmingham on Sunday 26 April 1987 in a blaze of Spring sunshine.

The procession of over 100 turnouts started off in the sylvan setting of Cannon Hill Park, and then wound its way along the surrounding roads, just up the road from Calthorpe Park, where it had last been held 35 years before.

A big crowd turned out to give a rousing welcome to the gaily decorated horses from huge Shires and Clydesdales, with feet as big as meat plates, to high-stepping Hackneys and costermongers' ponies. They had come from all over the country just for the pleasure of taking part.

A pair of superb and magnificently groomed greys pull the rubber-tyred wagon entered in the Birmingham parade by the Solid Fuel Advisory Service.

A sight to quicken the heart and gladden the eye . . . the climax of the 1987 parade as all the entrants salute the huge crowds whose support ensured the success of the event.

Leading the procession was a veteran double-decker horse omnibus, which carried the Lord Mayor of Birmingham, Councillor Denis Martineau and the Lady Mayoress, Mrs Mollie Martineau, and the civic party.

Also on board was the man whose single-minded initiative had brought back the old parade in all its glory.

For many years, Bryan Holden, chairman of a leading Birmingham advertising agency, had cherished the singular notion of reviving the Birmingham Horse Parade.

He lobbied civic departments, equestrian societies and the local media to build up support for the idea.

Eventually, he met with David Lewis, one of the foremost heavy horsemen in the country. Together, they gained the support of a handful of fellow horse enthusiasts and formed the Birmingham Harness Horse Society, with the express intention of reviving the old horse parade.

The City of Birmingham Leisure Services Committee, chaired by Councillor

Thoroughly enjoying a ride with a difference are the Lord Mayor, Councillor Denis Martineau, and the Lady Mayoress, on top of a vintage horse bus in the 1987 parade.

Brigadier Peter Fitchett leads his men of the Legion of Frontiersmen in a colourful display at the parade.

A pair of dapple greys provided the horse power for Birmingham City Market's stage coach entry in the big parade. And enhancing the occasion is a dapper gentleman in top hat and tails.

Bryan Bird, lent its wholehearted support, and the city's Recreational and Community Services Department helped in many practical ways to get the show on the road.

That the effort had been well worthwhile was endorsed by the warmth of public acclaim as the procession got underway, preceded by a ceremonial mounted police escort with scarlet plumes sweeping gracefully, like horses' tails, from their helmets.

Coachloads of senior citizens swelled the cheering throng. Some were in wheelchairs inside the main gates of the park, where they sat singing wartime songs that had kept spirits high in the bleak days of the Blitz.

Tears of joy were shed as the horses clattered by. There were cries of 'Well done!'. 'You've been away too long!'. 'Come back again next year'. 'God Bless you!'.

All along the 1½-mile route the excited people waved and shouted encouragement. A rousing cheer heralded the appearance of a posse of the mounted Legion of Frontiersmen in smart blue tunics and Canadian Mountie-like hats.

And there was more applause for Birmingham's most famous milkman, Albert Judd (Juddy) as he rattled by driving a horse provided by the Horse and Pony Rescue Association.

Horse omnibuses; coaches with top-hatted liveried coachmen; baker's carts; a barrel-top gypsy caravan; a pair of black funeral horses, with nodding black plumes between their ears, pulling a hearse; a replica 19th century London-Shrewsbury stage coach; brewers' drays, coal carts....

It was a horse-drawn panorama.A day for wallowing knee-deep in nostalgia - and for putting one's foot into rose grower's delight, scattered everywhere!

Not surprisingly, there were even more spectators the next time around -

The picture that says it all . . Albert Judd beams with delight as he manoeuvres his float through the admiring crowds at the 1987 Horse Parade in Cannon Hill Park.

Right:
'The Wonder,' this former London-Shrewsbury coach owned by parade administrator George Rawlins, proved a popular entry in the revived horse parade.

Far right:
Elegance and style are the hallmarks of these black beauties pulling a glass-sided hearse in the Birmingham parade.

Sid Rawlins, former Birmingham haulier and now living in retirement in Wales, needed no persuasion to return to the city to assess entries in the revived horse parade. Here he is at work.

A superb turn-out for Gallifords horsed by Brookfield Shires, of Northampton.

Mayoral congratulations for a premium award at the Birmingham parade.

It is a day for pride, perfection, and a little passion as the Birmingham Horse Parade is revived after many long years. This picture captures the spirit of the occasion - horses beautifully turned out, harness and brasses shining, and wagon sparkling. This was the Coalite entry at the 1987 parade.

in April 1988 - when some 30,000 packed into the park to see not only the horse parade, but also a day-long farriery competition.

In just three years the Birmingham Horse Parade has become a major national event supported by horsemen and horses from all over the country.

The parade attracts many of the finest turnouts driven by some of the best-known names in the business. But there is still a place for the amateur with his single horse and float - even for little Shetlanders pattering between the shafts of a governess cart.

Birmingham, for its part in this co-operative endeavour, has been presented

The farriery competition is a popular crowd-puller at the revived Birmingham Horse Parade. Britain's finest blacksmiths engage in a day long contest and the sound of hammer and anvil goes on from dawn to dusk.

It's not as easy as putting money in the bank and Mr. Ian Jones of sponsors Lloyds Bank (left), understandably needs help as he adorns the winner of the best black horse class with its prize sash. Bryan Holden is on the right.

The Birmingham Horse Parade is fully supported by Birmingham City Council, not least by Coun Bryan Bird, Chairman of Leisure Services, seen here admiring the fine hydrangeas decorating his department's entry.

with a national event which has raised the city's profile in other quarters than those influenced by Super Prix car racing.

'Birmingham can award itself a big rosette for trying and succeeding' said the Evening Mail, whose publicity helped to make people aware that the big parade was back.

The horses have left behind not only a literal whiff of a bygone era, but also the sweet smell of success.

The Birmingham Horse Parade has done much to reinforce the image of a vibrant, enterprising and imaginative Birmingham. A city with a glorious past, and one with an equally bright and challenging future.

For the joy of seeing a spectacle such as our grandparents had enjoyed we owe a debt of gratitude to Bryan Holden and the hard-working members of the Birmingham Harness Horse Society. ☐

Above:
Councillor Dick Knowles. leader of the Labour Group on Birmingham City Council, and his wife Ann acknowledge the cheers in the 1988 parade.

Above right:
His dream has become reality - no wonder Bryan Holden has one of the broadest smiles at the 1987 horse parade. With him is his wife Barbara.

The breweries, of course, played a prominent role in horse parades of the past - and they still maintain that interest. Here the crowds admire the Ansells Brewery entry at Cannon Hill Park.

BIRMINGHAM'S

14

Horses in the city

WORKING HORSES

David Lewis of Shire Shipping with Pride and Prince, two fine Shire geldings bound for an Australian brewery where, in 1986, they founded a new show team.

There is no likelihood that we shall ever see working horses in Birmingham, or anywhere else, for that matter, on a scale to challenge seriously the superiority of the motor vehicle.

Like its contemporary motive force of yesteryear, the steam locomotive, the working horse survives mainly because it is a revenue earning piece of living history. But that is not to say a horse can't hold its own against modern machinery. There are numerous

instances - organic farming, timber-snigging, short haul delivery, rubbish collection - all these are situations where flesh and blood horsepower has proved its economic viability.

That there is a place for working horses is recognised by the increase in the number of animals bred each year; and the growing membership of the four heavy horse breed societies.

In the early 1960s it seemed that the pedigree bloodline of the heavy horse might perish. No longer was there a market for heavies and only through the dedication of a handful of breeders did the Great Horse of England survive to blossom forth in the 1980s.

Younger Brummies could not believe their eyes and older ones had a sudden attack of nostalgia when they spotted this dustcart collecting rubbish in the city centre. It was part of a six-week trial held by the City Council in 1987. Holding the reins is Donald Knight, a member of the Veterinary Section.

Below right: Colin Whittock, whose cartoons appear daily in the Birmingham Evening Mail, let his imagination run riot on Derby Day 1987.

Below: And here's proof that a good idea won't lie down! A horsebus is again plying the streets of Birmingham but the year is 1988!

(Birmingham Daily News)

The pain of parting is evident in the faces of David and Barbara Lewis as they say goodbye to another of their superb stock. The picture was taken on the day Ladbrook Aristrocrat left for a new life in America.

One of the best known and respected stalwarts of the Shire horse was the late Arthur Lewis of Tanworth-in-Arden. His horses were counted among the survivors, but such was Mr Lewis's abilities as a horse breeder that they proved to be some of the finest of their type.

One of his best stallions was Ladbrook Invader. Standing 19.2 hands high to the top of its withers, Invader was at one time the world's biggest horse, a fact recorded in the Guinness Book of Records.

Eventually, he was sold to a leading Shire horse fancier in America.But the Little Ladbrook stud had another ace in the pack, and in 1980 Invader's son, Ladbrook Aristocrat, won the supreme championship at the National Shire Horse Show, Peterborough.

As with his sire, so Ladbrook Aristocrat is now in America, but many of his sons and daughters are doing well in Britain.

Ladbrook Courage, grandson of Invader, is a horse to follow in the championship show ring; and his patient, gentle nature has won the big fellow many admirers.

Ladbrook Courage is owned by David Lewis, Arthur Lewis's son. David and his wife Barbara run a horse export business and have pioneered the sea passage of Shires, and other pedigree breeds, by container vessel to Australia,

The Ladbrook Shire stud now have an international reputation for excellence and here is the daddy of them all - Ladbrook Invader. The outstanding stallion, exported to America, stands 19.2 hands high and was at one time the world's biggest horse. No wonder he dwarfs the cow-puncher at his new home in Maryland, Ohio.

where the demand for English bloodstock is high.

The Lewises have also saved Britain's last remaining firm making hand-crafted hames, and are training youngsters in the centuries-old craft.

Likewise, George Rawlins of Curdworth, was brought up with horses. His father, Sid Rawlins, headed a well-known Birmingham firm of hauliers and contractors.

But young George did not follow in the footprints of his father's Shire horses, rather he aspired to become a very accomplished whip in the age-old sport of driving.

George Rawlins's team of four-in-hand Gelderlanders is a stirring sight

The 'Shrewsbury Wonder' ran from London to Shrewsbury in 17 hours.
The coach belongs to George Wilkins, managing director, Pro-Mill Engineering, Tamworth.
Holding 'the ribbons' is George Rawlins, who supplied the team of Dutch Gelderlanders.
(Photo: Middleton Hall, Tamworth)

drawing a vintage road coach. They are familiar to many through their appearances at premier events all over the country, including the Royal Show.

Not only is George Rawlins a good judge of horseflesh, with educated hands that can drive a spirited team as though on a piece of cotton, he has an eye for vintage horse-drawn vehicles, too. George's opinion is much sought after when the 'Van Goghs' and 'Rembrandts' of the coaching world come under the auctioneers' hammer, at home and abroad.

As with David Lewis and George Rawlins, so Malcolm Tonks of Wythall has never known a time in his life without horses. His father was a greengrocer and supplied customers door-to-door. Young Malcolm got roped in as a delivery lad and was also expected to muck out the stables, groom the horses, and clean harness and brasses.

Fortunately, hard work didn't put him off, rather it sharpened his interest and stimulated a desire to have his own horses and work with them full-time.

For many years Malcolm Tonks was mine host at the Red Lion pub in Kings Heath. He remained in touch with horses, and made use of the old stables at the back of the pub. At weekends and bank holidays he was to be seen

Former publican Malcolm Tonks keeps a fine stable of horses and a collection of historic carriages at his farm in Wythall and is called on from time to time to present realistic looking scenes in films. Here he is in one such situation, riding his immaculate turnout over cobbled stones and defying any onlooker to claim that it does not look truly Victorian

You can find an equine Little and Large at Malcolm Tonks 'centre at Wythall. Here he is showing them off - Oddie the Shire and Pickles, a miniature Shetland pony.

immaculate in a brown bowler, riding breeches, and polished leather gaiters, driving blushing brides to church.

When Malcolm gave up the licenced trade it was to set up his own horse centre and throw himself wholly into equestrian work. His speciality is to provide film-makers and TV producers with authentic period pieces, using horses and vehicles, especially in historical dramas.

An indication of Malcolm's showmanship can be seen in his 'Little' and 'Large' equestrian twosome. He presents a full blown Shire gelding standing over 6 feet high (18 hands), alongside its stable-mate, Pickles, a miniature Shetland pony barely the size of an Alsatian dog. Together, they are a surefire show-stopper.

With expert horsemen such as these Birmingham can, indeed, have every confidence that the future of its horses is in safe and capable hands. □

Birmingham's Chief Veterinary Officer, Dr Bryan Davies, was given a new task and a new title for 'Queen Victoria's' charter centenary procession on 13 January 1989. He was 'Master of Horse'. Here he is checking one of the horses. But he did not ride in the parade, he walked behind to make sure everthing was going according to plan.
(Birmingham Post & Mail)